Praise for Marcia Bartusiak's *Black Hole:*

"Expertly tells the story of the emergence of black holes. . . . [Bartusiak] offers a concise but comprehensive history . . . from the 18th century ponderings of stars massive enough that light could not escape to present-day studies of these very real objects."—Jeff Foust, *Space Review*

"Lively and dramatic. . . . There's no danger of being bored. Bartusiak does a good job of tracing the twisted route that our understanding has followed, from Newton to Einstein and to today as we try to extend gravity to quantum scales."—Tara Shears, *Times Higher Education Supplement*

"*Black Hole* is engaging and lively, weaving in personal drama . . . with a clear account of the underlying science. An acclaimed science writer."—Tom Siegfried, *Science News*

"Superior science writing that eschews the usual fulsome biographies of eccentric geniuses, droll anecdotes and breathless prognostication to deliver a persistently fascinating portrait."—*Kirkus Reviews* (starred review)

"Lively and readable. . . . Read it if you want to know how the concept of black holes has changed dramatically over the past 100 years—from being an apparent mistake in the maths to the strangest and most outlandish objects that we (currently) know of."—Pippa Goldschmidt, *The Spectator*

"Tells this story with . . . an accessibility that other popular writers in the field have sometimes struggled to achieve when dealing with the subject. There's as much history and character study as hard science, which is no bad thing, and even the most mathophobic reader will find this book a satisfying and enlightening read."—Mike Parker, *Tribune Magazine*

"Bartusiak's lively, accessible writing and insight into the personalities behind the science make her book an entertaining and informative read."—*Publishers Weekly*

"Marcia Bartusiak's thorough, lucid biography of black holes shows how Albert Einstein's theory of general relativity posited how gravity produces these bizarre astrophysical objects."—Philip Ball, *Prospect*

"[A] reliable and readable account of this amazing story."—Andrew Crumey, *Literary Review*

"Astronomers took fifty years to carry the black hole from laughable concept to central importance in every galaxy. Marcia Bartusiak accomplishes the same feat here, in one irresistibly attractive read." —Dava Sobel, author of *Longitude*

"Marcia Bartusiak takes us on a fascinating ride around black holes, showing the beauty and mystery of a concept that has intrigued scientists from Einstein to Hawking."—Walter Isaacson, CEO of the Aspen Institute and author of *Benjamin Franklin, Einstein*, and *Steve Jobs*

"An engrossing and mind-bending read. . . . Bartusiak provides a front row seat as many of the most famous scientists of all time grapple with the strangest objects in the universe, black holes."—Adam Riess, Nobel Laureate in Physics, 2011

"Captivating and authoritative, *Black Hole* traces a truly weird concept from its tentative conjecture to inescapable reality. Bartusiak recounts a compelling tale with quirky turns, curious revelations, intellectual rumbles, and personal gambles."—Ray Jayawardhana, author of *Neutrino Hunters*

"Bartusiak's new book is thoroughly researched, beautifully written, and full of insights about the nature of the scientific enterprise. Aficionados of black holes will love this book."—Alan Lightman, author of *Einstein's Dreams* and *The Accidental Universe*

Black Hole

BLACK HOLE

HOW AN IDEA ABANDONED BY NEWTONIANS, HATED BY EINSTEIN, AND GAMBLED ON BY HAWKING BECAME LOVED

MARCIA BARTUSIAK

Yale

UNIVERSITY PRESS

New Haven & London

Yale University Press books may be purchased in quantity for educational,
business, or promotional use. For information, please e-mail sales.press@
yale.edu (US office) or sales@yaleup.co.uk (UK office).

Printed in the United States of America.

The Library of Congress has cataloged the hardcover edition as follows:
Bartusiak, Marcia, 1950–
Black hole: how an idea abandoned by Newtonians, hated by Einstein,
and gambled on by Hawking became loved / Marcia Bartusiak.
pages   cm.
Includes bibliographical references and index.
ISBN 978-0-300-21085-9 (clothbound : alk. paper)
1. Black holes (Astronomy).   2. Discoveries in science.
3. Science—Social aspects. I. Title.
QB843.B55B37 2015
523.8'875—dc23   2014038950

ISBN 978-0-300-21966-1 (pbk.)

A catalogue record for this book is available from the British Library.

10 9 8 7 6 5 4 3 2 1

Dedicated to my students, past and present, in the Graduate Program in Science Writing at the Massachusetts Institute of Technology, who daily inspire me

Contents

CONTENTS

Preface

The very notion of a black hole is so alluring. It combines the thrill of the unknown with a sense of lurking danger and abandon. To imagine a journey to a black hole's outer boundary is like approaching the precipice of Niagara Falls, contemplating the vertical plunge to the churning waters below, yet remaining secure in the knowledge that we're positioned behind a sturdy fence to keep us from peril. Even in the real world, we know we're safe, as the closest black holes to Earth are thankfully hundreds of light-years distant. So, we experience the dark, celestial adventure vicariously.

For any astrophysicist at a cocktail party, it's the cosmic object they're most likely to be asked about. And for good reason: a black hole is wackily weird. As noted black-hole expert and Caltech theorist Kip Thorne has written, "Like unicorns and gargoyles, black holes seem more at home in the realms of science fiction and ancient myth than in the real Universe."

University of Texas astrophysicist J. Craig Wheeler calls them a cultural icon. "Nearly everyone understands the symbolism of black holes as yawning maws that swallow everything and let nothing out," he says.

And it was the same stark and alien weirdness, now celebrated, that kept physicists from accepting black holes for decades on end. According to a famous saying, often quoted, "All truth passes through three stages: First, it is ridiculed; second, it is violently opposed; and third, it is accepted as being self-evident." The concept of the black hole fully experienced each and every phase.

It's the black hole that forced both astronomers and physicists to take Albert Einstein's most notable achievement—general relativity—seriously. For a time the theory had entered a valley of despair. Einstein was honored as the "Person of the Twentieth Century" by *Time* magazine, yet such an honor would have been a huge surprise to the scientific community in midcentury. In that era, few universities in the world even taught general relativity, believing it had no practical applications for physicists. The best and the brightest flocked to other realms of physics. After the flurry of excitement in 1919, when a famous solar eclipse measurement triumphantly provided the proof for Einstein's general theory of relativity, the noted physicist's new outlook on gravity came to be largely ignored. Isaac Newton's take on gravity worked just fine in our everyday world of low velocities and normal stars, so why be concerned with the minuscule adjustments that general relativity offered? What was its use? "Einstein's predictions refer to such minute departures from the Newtonian theory," noted one critic, "that I do not see what all the fuss is about." After a while, Einstein's revised vision of gravity appeared to have no particular relevance at all. By the time that Einstein died, in 1955, general relativity was in the doldrums. Only a handful of physicists were specializing in the field. As Nobel laureate Max Born, a longstanding and intimate friend of Einstein, confessed in a conference the year of Einstein's death, general relativity "appealed to me like a work of art, to be enjoyed and admired from a distance."

But in reality what Einstein had done was devise a theory that was decades ahead of its time. Experimental measurements had to catch up to his model of gravity, which had been fashioned from pure intuitive thought. Not until astronomers revealed surprising new phenomena in the universe, brought about with advanced technologies, did scientists take a second and more serious look at Einstein's view of gravity. Observers identified in 1963 the first quasar, a remote young galaxy disgorging the energy of a trillion suns from its center. Four years later, much closer to home, observers stumbled upon the first pulsar, a rapidly spinning beacon emitting staccato radio beeps. Meanwhile space-borne sensors spotted powerful X-rays and gamma rays streaming from points around the celestial sky. All these new and bewildering signals pinpointed collapsed stellar objects—neutron stars and black holes—whose crushing gravity and dizzying spins turn them into extraordinary dynamos. With the detection of these new objects, the once sedate universe took on a racy edge; it metamorphosed into an Einsteinian cosmos, filled with sources of titanic energies that can be understood only in the light of relativity.

And what astrophysicists ultimately discovered and came to appreciate was general relativity's deeper aesthetic appeal, especially when it came to black holes. "They are," said Subrahmanyan Chandrasekhar on receiving the Nobel Prize in physics in 1983, "the most perfect macroscopic objects there are in the universe." Black holes offered all that a physicist yearned for in a theoretical outcome: both simplicity and beauty. "Beauty," Chandrasekhar told the audience, "is the splendor of truth."

Where once the field of general relativity was a cozy backwater, it now flourishes, both in theory and in practice. The black hole is no longer an oddity but a vital component of the universe. Nearly every fully developed galaxy appears to have a supermassive black hole at its center; it may be that the very existence of a galaxy depends on it. Telescopes are currently closing in on the gargantuan hole that resides in the heart of our home galaxy, the Milky Way. At the same time, cutting-edge observatories, newly designed, stand ready to detect the space-time rumbles—gravity waves—emitted when black holes collide somewhere in our intergalactic neighborhood. As John Archibald Wheeler, a dean of American relativity, once noted in the dedication of his autobiography, "We will first understand how simple the universe is when we recognize how strange it is."

But arriving at that knowledge took more than two hundred years—from the precursor of the black-hole idea in the 1780s to observational proof in the latter half of the twentieth century. And for most of that time, the very notion of this strange entity's presence in the cosmos was either ignored or earnestly fought against. Physicists conceded the black hole's existence only after a lot of virtual kicking and screaming.

In hindsight, it's difficult to understand why they put up such a struggle. The idea of the black hole is actually quite simple. It has a mass, and it has a spin. In some ways, it's as elementary an entity as an electron or a quark. But what confounded physicists for so long was the black hole's ultimate nature: it's matter squeezed to a point. Their fight against such an outcome for a star was more philosophy than science; they deeply believed nature wouldn't (couldn't!) act in such a crazy way. It took a handful of physicists over that lost half century to swim against the current and push the idea forward, crazy or not. Now, in celebration of general relativity's hundredth anniversary in 2015, here is the story of that frustrating, resourceful, exhilarating, and (at times) humorous struggle toward acceptance. This is neither the anatomy of a black hole nor a report on the latest astronomical and theoretical findings. This is the history of an idea.

Black Hole

1

It Is Therefore Possible That the Largest Luminous Bodies in the Universe May Be Invisible

It all began with Sir Isaac Newton.

No, let me take that back. The ancestry of the black hole can actually be traced back much earlier. You could say it began in ancient times when that distant era's most clever thinkers—the now-forgotten Newtons and Einsteins of their day—pondered why our feet stay firmly planted on the ground. That was an obvious question for a budding scholar to ask in that bygone age.

Everything centers on gravity. Gravity controls the sweep of the planets around the Sun, as well as the fall of a fluttering leaf from an autumnal tree. It is a force we take for granted but took us centuries to understand. Why is stuff attracted downward, toward the surface of the Earth? More than two thousand years ago, Aristotle and other ancient philosophers had a ready and reasonable reply to that question: our planet resided at the very center of the universe, so naturally everything fell toward it. Humans, horses, carts, and buckets were all driven to reside at the prime position. As a consequence, we were sturdily attached to terra firma. It was the natural state of things.

This explanation made perfect sense, given our daily experience. That is, until Nicolaus Copernicus stepped in and dramatically changed the cosmic landscape once and for all. In 1543 the Polish administrative canon dared to assert that Earth was in reality orbiting the Sun with all the other planets. Others, such as Aristarchus of Samos in the third century BCE, had suggested this scheme before, but not until Copernicus introduced his heliocentric universe did it finally take root. As a result, the long-assumed setup that kept us bound to our planet had to be completely reexamined. Earth was no longer at rest in the universe's hub, serenely waiting for objects to rain down upon it. Instead, the Earth was yanked into motion and the Sun was now at center stage. This new alignment soon motivated some of the greatest minds in Europe to reassess the rules of gravity, as well as the mechanism behind planetary motion. The challenge was on.

Inspired by Englishman William Gilbert's claim in 1600 that the Earth was a giant magnet, the German mathematician Johannes Kepler suggested that threads of magnetic force emanating from the Sun were responsible for pushing the planets around. The French philosopher René Descartes in the 1630s, in contrast, imagined that the planets were carried around like leaves trapped within a swirling whirlpool by vortices of *aether,* the tenuous substance then thought to permeate the universe.

All these ideas would eventually be overturned, though, once Isaac Newton in 1687 offered a more rigorous set of rules for both gravity and planetary motion. That was the year he published his masterful *Philosophiae Naturalis Principia Mathematica* (Mathematical principles of natural philosophy). We know it today as simply the *Principia.* Newton was forty-four years of age at the time, but his new take on gravity had been percolating in his mind long before that.

It began in 1665 during the Restoration reign of King Charles II, when the black plague was once again rampant. To wait out the epidemic, Newton temporarily left his studies at Cambridge University and retreated to his childhood home in the rural hamlet of Woolsthorpe, just east of Nottingham. It was there that the precocious student possibly watched that fabled apple fall in his country garden, which inspired him to reflect on the tendency of bodies to fall toward the Earth with a set acceleration. Did the same force acting on the apple extend to the Moon, he asked himself? A virtuoso at mathematics, largely self-taught, he computed that the Moon did seem to be continually "falling" toward the Earth—its path becoming curved—by an earthly pull that diminished outward by about the square of the distance. In other words, double the distance between two objects and the force between them is reduced to one-fourth of its original strength. Triple the distance and the force is diminished to one-ninth. Mathematically, this is a sign that a force is spreading its influence equally in all directions. But since Newton's early figurings weren't flawless, the young man put the problem aside for many years. "He hesitated and floundered," wrote Newton biographer Richard Westfall, "baffled for the moment by overwhelming complexities."

Newton's interest in gravitation didn't fully resurface until the 1670s. That was the period when Robert Hooke, curator of experiments for Great Britain's Royal Society, developed an appealing set of conjectures for explaining gravitation: that all celestial bodies have a gravitating power directed toward their centers; that they can attract other bodies; and that the attraction is stronger the closer you are to the body. Hooke had a general set of rules but as yet no equations. What he couldn't figure out, as he noted in his published paper, was whether planetary motion would necessarily be "a Circle, Ellipsis, or some other more compounded Curve Line." Triggered by an exchange of letters with Hooke on this topic in the winter of 1679–80, Newton was galvanized to return to the problem of his youth.

The famous tree (center) at Woolsthorpe Manor, England, from which Isaac Newton allegedly saw an apple fall under the influence of gravity. (*Roy Bishop, Acadia University, courtesy of American Institute of Physics Emilio Segrè Visual Archives*)

Yet, at first, he kept his revolutionary results to himself. That was because Newton, an intensely private man, was wary of his jealous rival Hooke. Often fearful of exposing his work to criticism, he once confessed in a letter to a colleague, "I am . . . shy of setting pen to paper about anything that may lead into disputes." We largely have Edmond Halley (the famous comet's namesake) to thank for

Newton's writing the *Principia* at all. While visiting in 1684, Halley asked the illustrious physicist how a planet would move under an inverse square law. Newton confidently replied, "An ellipsis," what we now call an ellipse, noting that he had worked it out several years before.

From that very moment, Halley became Newton's staunchest advocate. It was Halley's persistent prodding and financial backing that finally energized Newton to write his masterpiece on gravity. And once committed, Newton didn't hold back. Westfall notes that Newton had an enormous capacity for "ecstasy, total surrender to a commanding interest," often forgetting to eat or sleep when in this state. Halley had now sparked that intellectual rapture once again. Newton swiftly abandoned his ongoing projects (among them, classical mathematics, theology, and alchemy) and fully applied his legendary power of concentration on completing his work on gravity. Armed with better measurements of the Earth, he was at last able to prove decisively that an inverse square law attracted the Moon to the Earth and that such a force directly leads to planets moving in elliptical orbits, just as Kepler had revealed in 1609. Kepler knew from his measurements that planets moved in elliptical orbits but did not know the reason. Decades later, Newton showed through his mathematics that such orbits were a natural consequence of the law of gravity. Observation and theory, working from opposite corners, came together and matched.

It took Newton nearly two years to complete the *Principia,* and for understandable reasons. Encouraged by his initial and successful calculations, Newton worked out more and more cases with his new rules, and as a result long-standing problems in astronomy seemed to dissolve away. Gravity could now explain the tides and the Earth's precession (due to the Moon and Sun's tugging on the Earth's bulge), as well as the trajectory of a comet. In a grand conjectural leap,

Newton was declaring that gravity was a fundamental and universal force of nature. That term *universal* was a key insight. What draws an apple to the ground also keeps the Moon in orbit about the Earth. "For nature is simple," Newton wrote, "and does not indulge in the luxury of superfluous causes." The cosmos and terra firma were no longer separate realms, as Aristotle had long reasoned; the heavens and the Earth now operated under *one* set of physical laws. Gravity, the attraction of one body for another, acts in a similar manner on all levels of the cosmos—on Earth, within the solar system, as well as among the stars, galaxies, and clusters of galaxies.

There was one problem with Newton's law of gravity, though. It implied that imperceptible ribbons of attraction somehow radiated over distances, both short and long, to draw moon to planet and boulder to Earth. For many, this feat appeared more resonant with mysticism than science. Critics demanded a physical mechanism. That is what natural philosophers had been providing for centuries. In what way was gravity doing its work? What was replacing either magnetism or vortices? This led to Newton's famous statement in the *Principia:* "I have not as yet been able to deduce from phenomena the reason for these properties of gravity, and I do not feign hypotheses." Newton, unlike his contemporaries, was not going to stoop to speculating or conjuring up some kind of hidden cosmic machinery. He was essentially satisfied that his laws allowed physicists to calculate, with great accuracy, the movement of a planet or the path of a cannonball. As the years passed, the rest of the physics community eventually came over to Newton's side. What greatly helped was the visit of a celestial traveler.

After poring over historic records, Halley figured that a comet sighted in 1682 had much in common with comets previously observed in 1531 and 1607. They shared the same orbital characteristics, going around the Sun in the opposite direction to the planets, and appeared every seventy-five to seventy-six years. Upon calculating the comet's orbit based on Newton's laws, he predicted in 1705 that the comet would return at the end of 1758. And so it did, right on schedule sixteen years after Halley's death, thirty-one years after Newton's. This feat bedazzled and instantly silenced Newton's critics. Who could argue with a theory that allowed for a spot-on prediction about the solar system's behavior more than half a century in advance? It was at that moment that Newton's law of gravity, despite its lack of a mechanism, was at last victorious.

With Newton's laws in place, scientists in the eighteenth century came to view the universe as intrinsically knowable, ticking away like a well-oiled timepiece. Many astronomers began spending long hours huddled at their desks using Newton's mathematical rules to work out planetary motions and to forecast the tides. Stars, as well, became convenient objects for testing the laws of gravity. And it was during such a stellar endeavor that a precursor to the black hole—the Model-T version, in a way—emerged. The possible existence of such a bizarre object arose when an Englishman named John Michell applied Newton's laws to the most extreme case imaginable.

Michell stood in the very thick of a wondrous age of scientific discovery, and he dabbled in it all. He was a geologist, astronomer, mathematician, and theorist who regularly hobnobbed with the greats of the Royal Society of London, such men as Henry Cavendish, Joseph Priestley, and even the society's American fellow Benjamin Franklin (during the diplomat's two long stays in London). The claim could be made, science historian Russell McCormmach has written, that Michell was "the most inventive of the eighteenth-century natural philosophers." He recognized early on, for example, that Earth's strata could bend, fold, rise, and dip. If Michell is remembered at all today, it is for his suggestion in 1760 that earthquakes propagate as elastic waves through the Earth's crust. That earned Michell the title "father of modern seismology." By deeply scrutinizing and comparing various accounts of the great temblor that leveled Lisbon, Portugal, in 1755, Michell was able to compute the time, location, and depth of the quake's epicenter, situated to the west in the Atlantic Ocean.

Michell also designed a delicate instrument that could measure the gravitational constant in Newton's equations, allowing him in essence to "weigh" the Earth. He died before he could carry out the experiment, but his friend Cavendish ultimately obtained the torsion balance, made modifications, and successfully measured our planet's mass with it.

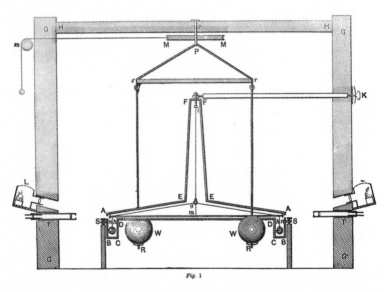

Fig. 1

The torsion balance, based on a design by John Michell, that was used by Henry Cavendish in 1797–98 to weigh the Earth. (*Philosophical Transactions of the Royal Society of London*)

Despite such accomplishments, however, Michell had the unfortunate habit of burying original insights (such as the inverse-square law of magnetic force decades before it was rediscovered) in journal papers that focused on more mundane research, and so received little notice. Some of his greatest ideas were casually mentioned in asides or footnotes. As a consequence, long-lasting fame eluded him.

Michell had begun his scientific investigations at Queens' College in Cambridge. Son of an Anglican rector, he entered Queens' in 1742 at the age of seventeen and after graduation remained there to teach for many years. A contemporary described him as a "short Man, of a black Complexion, and fat. . . . He was esteemed a very ingenious Man, and an excellent Philosopher." While in Cambridge, Michell even tutored a young Erasmus Darwin, Charles Darwin's grandfather, who called his mentor "a comet of the first magnitude." .

But by 1763, ready to marry, Michell decided to leave teaching and devote himself to the church. He ultimately settled in the village of Thornhill in West Yorkshire, where he served as a clergyman until his death in 1793 at the age of sixty-eight. Yet, over those decades with the Church of England, the reverend continued to indulge his wide-ranging scientific curiosity. He had a nose for interesting questions and was willing to stick his neck out in speculation, though always grounded in his first-rate mathematical skills. One of Michell's more intriguing conjectures at this time, right when Great Britain was recovering from its war with colonial America, was imagining what today we call a black hole.

This idea grew out of an earlier prediction that Michell had made. Astronomers in the eighteenth century were starting to see more and more double stars as they scanned the celestial sky with their ever-improving telescopes. The common wisdom of the time declared that such stars were actually at varying distances from Earth and closely aligned in the sky by chance alone—that it was just an illusion that they were connected in any way. But, with remarkable insight, Michell argued that nearly all those doubles had to be gravitationally bound together.

He was suggesting that some stars exist in pairs, a completely novel notion for astronomers in those days. In a groundbreaking paper published in 1767, Michell worked out the high probability that, given how most other stars were arranged on the sky, the twin stars were physically near each other—"the odds against the contrary opinion," he stressed, "being many million millions to one." (As usual, he displayed his computations in a footnote.) In carrying out this calculation, Michell was the first person to add statistics to astronomy's repertoire of mathematical tools. This paper, according to astronomy historian Michael Hoskin, was "arguably the most innovative and perceptive contribution to stellar astronomy . . . in the eighteenth century."

At the same time, Michell recognized that double stars would be quite handy for learning lots of good things about the properties of stars—how bright they are, how much they weigh, how vast is their girth. Michell suspected that there were stars both brighter and dimmer than our Sun. He cunningly ventured that a white star was brighter than a red one. ("Those fires, which produce the whitest light," he pointed out in his paper, "are much the brightest.") So, two stars orbiting each other were the perfect laboratory for testing his ideas from afar and arriving at answers. Yet, nearly all astronomers in his day weren't concerned with such questions. They were too busy discovering new planetary moons or tracking the motions of the planets with exquisite precision. To them, the stars themselves were not terribly interesting and served merely as a convenient backdrop for their measurements of the solar system and its components. The

Sun, Moon, and planets were astronomers' prime observing targets at the time.

The great British astronomer William Herschel, a friend of Michell's, was the rare exception to that rule. He often veered away from conventional astronomical work. Within a dozen years of Michell's paper on double stars, Herschel began monitoring and cataloging the stars positioned close together in the sky. Michell extolled Herschel's growing data bank as "a most valuable present to the astronomical world." So valuable that Michell extended his ideas on double stars in a paper, published in 1784, with the marathonic title "On the Means of discovering the Distance, Magnitude, &c. of the Fixed Stars, in consequence of the Diminution of the Velocity of their Light, in case such a Diminution should be found to take place in any of them, and such other Data should be procured from Observations, as would be farther necessary for that Purpose." (*Whew!*) It was in this work that Michell hinted at the possibility of a black hole—or at least an eighteenth-century, Newtonian version of one.

The eminent Henry Cavendish, discoverer of hydrogen and its connection to water, read Michell's paper before the Royal Society over a succession of meetings, in November and December 1783 as well as January 1784. (It was then published in the Royal Society's *Philosophical Transactions,* taking up twenty-three pages in print.) Michell was devoted to the society and at least once a year traveled the arduous two hundred miles (three hundred kilometers) from Yorkshire to London to either attend its meetings or meet with society friends. But for those winter meetings the reverend inexplicably stayed home. It could have been ill health or lack of travel funds, or possibly he just wanted to avoid a raucous skirmish then in full swing to unseat the society's president, the botanist Sir Joseph Banks. There was also the matter of preliminary tests of his idea. They weren't

VII. *On the Means of difcovering the Diftance, Magnitude,* &c.
of the Fixed Stars, in confequence of the Diminution of the
Velocity of their Light, in cafe fuch a Diminution fhould be
found to take place in any of them, and fuch other Data fhould be
procured from Obfervations, as would be farther neceffary for
that Purpofe. By the Rev. John Michell, *B. D. F. R. S.*
In a Letter to Henry Cavendifh, *Efq. F. R. S. and A. S.*

Read November 27, 1783.

DEAR SIR,   Thornhill, May 26, 1783.

THE method, which I mentioned to you when I was laft
in London, by which it might perhaps be poffible to
find the diftance, magnitude, and weight of fome of the fixed
ftars, by means of the diminution of the velocity of their
light, occurred to me foon after I wrote what is mentioned by
Dr. PRIESTLEY in his Hiftory of Optics, concerning the diminution of the velocity of light in confequence of the attraction of the fun ; but the extreme difficulty, and perhaps im-
poffibility, of procuring the other data neceffary for this purpofe appeared to me to be fuch objections againft the fcheme,
when I firft thought of it, that I gave it then no farther confi-
deration. As fome late obfervations, however, begin to give
us a little more chance of procuring fome at leaft of thefe data,
I thought it would not be amifs, that aftronomers fhould be
apprized of the method, I propofe (which, as far as I know,
F 2   has.

The eighteenth-century scientific paper in which John
Michell first suggested the existence of the Newtonian
version of a "black hole." (*Philosophical Transactions of
the Royal Society of London*)

detecting what he had hoped to measure. But some historians have
speculated that Michell recognized the daring nature of his paper and
thought the society would more readily accept his idea if his close
friend and highly respected colleague presented it.

The radical technique that Michell was proposing to study the
stars involved the speed of light. If astronomers closely monitored the
two stars in a binary system moving around each other over the years,

noted Michell, they could calculate the masses of the stars. It was the most basic application of Newton's laws of gravity. If he measured the width of the orbit and the time it takes for the two stars to orbit each other, he could estimate their masses. And if each star's gravitational pull affected the other's motion, suggested Michell, that pull should also affect light. This was an era when light was assumed to be composed of "corpuscles," swarms of particles—largely because Newton, whose opinion was revered, had endorsed that idea.

Now imagine these particles journeying off a star and out into space. Michell assumed that gravity would attract these corpuscles just like matter. The bigger the star, the stronger the gravitational hold upon the light, slowing down its speed. There would be a "diminution of the velocity of [the stars'] light," as the title of his paper announced. Measure the velocity of a beam of starlight entering a telescope, and voilà, you acquire a means of weighing the star.

Now, this is where the "black hole" possibility arises: Michell took his scenario to the utmost limit and estimated when the mass of the star would be so great that "all light . . . would be made to return towards it, by its own proper gravity"—like a spray of water shooting up from a fountain, reaching a maximum height, and then plunging back down to the bowl. With not one radiant corpuscle escaping from the star, it would remain forever invisible, like a dark pinpoint upon the sky. According to Michell's calculations, this transformation would occur when the star was about five hundred times wider than our Sun and just as dense throughout. In our solar system, such a star would extend past the orbit of Mars.

In 1796, in the midst of the French Revolution, the mathematician Pierre-Simon de Laplace independently arrived at a similar conclusion. He briefly mentioned these *corps obscurs,* or hidden bodies, in his famous *Exposition du système du monde* (The system of the world), essentially a handbook on the cosmology of his day. "A luminous star, of the same density as the Earth, and whose diameter should be two hundred and fifty times larger than that of the Sun," he wrote, "would not, in consequence of its attraction, allow any of its rays to arrive at us; it is therefore possible that the largest luminous bodies in the universe, may, through this cause, be invisible." It was only after an appeal from a dogged colleague, the astronomer Baron Franz Xaver von Zach, that Laplace three years later worked out a rigorous mathematical proof to back up his initial, cursory statement. Laplace's estimate for the width of the dark star differed from Michell's because he assumed a greater density for sunlike bodies.

But did it even make sense to predict the existence of stars that could never be seen? Laplace may have had second thoughts when light came to be viewed as waves, not corpuscles. Or perhaps he simply experienced a loss of interest, for in subsequent editions of *Système du monde,* which he published up until his death in 1827, he expunged his invisible-star speculation and never referred to it again. Michell, in contrast, displayed greater ingenuity in his 1784 paper. There he suggested a clever way to "see" such invisible stars. If one revolved around a luminous star, he noted, its gravitational effect on the bright star's motions would be noticeable. In other words, the bright star would appear over time to jiggle back and forth on the sky, due to the dark star's tugs. It's one of the very ways that astronomers today track down black holes.

In the end, though, Michell and Laplace were getting ahead of themselves—contemplating problems before the physics was in place to answer them. They didn't yet realize that supergiant stars have far lower densities than the ones they envisioned. They also never considered that the same invisibility effect could happen if a star were smaller but very, very dense. If an ordinary star were somehow compressed into a smaller volume, the velocity needed to escape from its surface would increase appreciably. But astronomers then just assumed that all stars shared the same density as the Sun or Earth. Could anything be denser than the elements found on Earth? It seemed unthinkable in the late eighteenth century.

Both Michell and Laplace were working with an inadequate law of gravity and the wrong theory of light. They didn't yet know that light never slows down in empty space. Proving the existence of such dark stars required more advanced theories of light, gravity, and matter. The modern conception of the black hole—a real "hole" in space-time, rather than just a big, dark lump of stellar matter—would not emerge for nearly a century. It had to wait for the entrance of the twentieth century's most inventive natural philosopher, Albert Einstein.

2

Newton, Forgive Me

Physicists could proudly point to two major accomplishments by the end of the nineteenth century: Newton's classical mechanics (established more than two centuries earlier) and the equations of electromagnetism formulated by the Scottish theoretician James Clerk Maxwell in the 1860s. In the physical sciences, each was the monumental theory of its age. It was Maxwell who figured out what light truly was, linking it to the long-known phenomena of electricity and magnetism. He predicted the existence of electromagnetic waves, whose "velocity is so nearly that of light," he reported, "that it seems we have strong reason to conclude that light itself . . . is an electromagnetic disturbance in the form of waves." And from his equations alone, Maxwell had revealed a new, fundamental constant in nature—the speed of light.

Because the scientific principles set forth by both Newton and Maxwell yielded extremely accurate predictions in experiment after experiment, it was easy to think that little remained to be done on these topics. Even Maxwell in 1874, during an address at the opening of what is now the Cavendish Physical Laboratory at Cambridge University, wondered if "the only occupation which will . . . be left to men of science will be to carry on these measurements to another place of decimals."

But to an inquisitive German student of physics in the 1890s, there was something not quite right about these laws. Taken together, they seemed somewhat out of kilter in his mind. His suspicions involved many complexities, such as the true nature of the ether believed to fill space and how light traveled through it. But, at its core, what disturbed the young Albert Einstein was that these two great works of physics didn't appear to share the same rules on handling space and time. Fearless at challenging the greats of his day, even as a student, Einstein was sure that the prevailing theory linking Newton's mechanics with Maxwell's electromagnetism—electrodynamics—was "not correct, and that it should be possible to present it in a simpler way." He yearned to make the two theories fully compatible with one another.

This wasn't a sudden decision on his part. The root of this challenge went back to his adolescence. In some autobiographical notes, Einstein recalled being caught up in a particular reverie: If a man could keep pace with a beam of light, what would he see? Would he observe an electromagnetic wave frozen in place, like some glacial swell? "There seems to be no such thing," he remembered thinking, at the youthful age of sixteen. According to Newton, you could catch up to the light, much like two runners in a relay race. But from Maxwell's perspective, that wasn't so clear. Experiments measuring how fast light traveled through the "ether" were suggesting that there was no catching up.

After spending years, on and off, thinking about this issue, Einstein at last arrived at a solution. But more than that, he did so by making the most basic assumptions possible. No grand leaps of theoretical prowess were required to reach an answer. Einstein's historic 1905 paper—on what came to be called special relativity—is actually exquisite in its simplicity. All of his hypotheses are based on physics that was available to anyone in the nineteenth or early twentieth century. Indeed, the same concerns bedeviled other physicists who were close to a solution, but all kept missing the key ingredient. Einstein's one inventive assumption was an entirely new conception of space and time. With that single change, all fell into place; the mismatch between Newton and Maxwell vanished.

Special relativity proposed that all the laws of physics (for both mechanics and electromagnetic processes) are the same for two frames of reference: one at rest and one moving at a constant velocity. Those schooled in Newtonian physics already knew that a ball thrown upward on a train moving forward at a steady 100 miles (160 kilometers) per hour behaves exactly the same as a ball thrown into the air from a motionless playground. Einstein wanted that agreement to be true for electromagnetism as well. But that meant that the behavior of a beam of light—that is, its measured velocity of 186,282 miles (299,792 kilometers) per second—must be the same in each place, both on the speedy train and on the ground. Why? If the laws of physics are to remain the same in both settings—on both the steadily moving train and the playground—the speed of light must be identical in both environments. "[We will] introduce another postulate . . . ," wrote Einstein in his 1905 paper, "that light is always propagated in empty space with a definite velocity c which is independent of the state of motion of the emitting body."

That seems like a reasonable assumption, until you make the comparison fairly drastic. The effects of Einstein's assertion are not really noticeable unless the comparative speeds are extremely high. So, let's do that: Consider a spaceship consistently moving away from Earth at 185,000 miles per second, just under the velocity of light. Common sense might lead you to believe that the astronauts would be going nearly as fast as any light beam passing by the Earth—that they even had a chance of overtaking the light, if they went a little faster, as Einstein once contemplated. But that's not the case at all.

The astronauts on that spaceship will still measure the velocity of that passing light beam at 186,282 miles per second, just as we do back here on Earth.

This situation seems bizarre, but only because our commonplace notions of space and time get in the way. In our everyday lives we think like Newton and all the ancient philosophers before him did: that space is an empty box, forever the same and immovable. You are either at rest or in motion within this fixed space that surrounds us.

Similarly, there is a universal clock, which ticks off the seconds for all the inhabitants of the cosmos in the same way. Events everywhere, from one end of the universe to the other, are in step with this grand cosmic timepiece, no matter what your position or speed.

But Einstein ingeniously realized that wasn't the case at all. The seeming paradox that arises for those fast-moving astronauts—how they can possibly measure the same speed of light as us—is solved by acknowledging that time is not absolute. Time is, well, relative. The very term *velocity* (miles per hour or meters per second) involves keeping track of time, but the astronauts and earthlings do not share the same time standard. That was Einstein's genius. He recognized that Newton's universal clock was a sham.

Since nothing can travel faster in a vacuum than the speed of light, two observers set apart in different frames of reference cannot agree on what time it is. The finite speed of light prevents the two from simultaneously synchronizing their watches. Einstein discovered that observers separated by distance and movement will not agree on when events in the universe are taking place.

This mismatch has other consequences as well. By just looking, the earthlings and astronauts will also not agree on one another's measurements. Mass, length, and time are all adjustable, depending on one's individual frame of reference. Look from Earth at a clock on that swiftly receding spaceship. You will see time progressing more slowly than here on Earth. You will also see the spaceship foreshortened in the direction of its motion. Those on the spaceship, who perceive no changes in themselves or in their clock's progression, look back at their receding home planet and see the same contraction of objects and slowing of time in the earthlings' surroundings! Each of us measures a difference in the other to the same degree. Space shrinks and time slows down when two observers are uniformly speeding either toward or away from each other. Space and time are different in each reference frame—enough to keep the speed of light the same in both environments. As soon as the astronauts start moving in relation to the Earth, they spawn (in a sense) their own "bubble" of experience, different from ours. We no longer share the same worldview. The only thing that we earthlings and the astronauts will agree on is the speed of light in a vacuum. That is the one universal constant.

With absolute time destroyed, there was also no need for absolute space either. Our intuition that the solar system sits serenely at rest, with the spaceship speeding away in some motionless container of space, no longer works. No such entity exists. In reality, we could just as easily consider the astronauts at rest, with the Earth speeding away. That being the case, wrote Einstein, the ether becomes "superfluous." Armed with this new viewpoint, he said, physicists no longer needed "an 'absolutely stationary space' provided with special properties." The ether had once provided a unique reference frame for physicists; it marked an absolute and universal state of rest. But Einstein revealed that this ethereal substance had been a fiction all along. "For me— and many others—the exciting feature of this paper was not so much its simplicity and completeness," said physicist Max Born during a fiftieth-anniversary celebration of the publication of special relativity, "but the audacity to challenge Isaac Newton's established philosophy, the traditional concepts of space and time."

The mathematician Hermann Minkowski, who once taught Einstein, brilliantly discerned an even deeper beauty in Einstein's new theory. With his expert mathematical know-how, Minkowski recognized that he could recast special relativity into a geometric model. He showed that Einstein was essentially making time a fourth dimension. Space and time coalesce into a single entity known as space-time. You can think of space-time as a series of snapshots stacked together, tracing changes in space over the seconds, minutes, and hours. Only now the snapshots are melded together into an unbreakable whole. Dimensionally, time acts like just another component of space. It's a consequence of the speed of light being constant; speed is defined as distance over time. If the distance a light beam travels contracts, so too must time slow down, to keep the ratio a constant. The two are irrevocably linked. "Henceforth," said Minkowski in a famous 1908 lecture, "space by itself, and time by itself, are doomed to fade away into mere shadows, and only a kind of union of the two will preserve an independent reality."

Minkowski cleverly recognized that, although different observers in different situations may disagree on when and where an event occurred, they will agree on a combination of the two. From one position, an observer will measure a certain distance and time interval between two events. Perched in another frame of reference, a different observer may see more space or less time. But in both cases they will see that the *total* space-time separation is the same. The fundamental quantity becomes not space alone, or time alone, but rather a combination of all four dimensions at once—height, width, breadth, and time. Einstein was not an expert in mathematics and so didn't appreciate this geometrical representation. When first acquainted with

Minkowski's idea, he declared the abstract mathematical formulation "banal" and "a superfluous learnedness." This was because Minkowski's novel outlook, to him, didn't seem to offer any added value to the physics he had so meticulously constructed. But he would soon change his mind.

Special relativity is called "special" for a reason. It covers only a very restricted type of motion: objects moving straight ahead at a constant velocity. That's a pretty narrow range of movement. So, soon after devising this new law, Einstein was determined to extend its rules to *all* types of motion, things that are accelerating away from us, slowing down, twisting, or turning. But special relativity was "child's play," said Einstein, compared to the development of a more *general* theory of relativity, one that would cover all those additional dynamical situations—in particular, gravity, which involves acceleration.

Over the ensuing years Einstein's reputation grew and soared, which some of his former teachers found surprising. Often bored in class, he had antagonized his professors, which made it difficult for him to obtain an academic position upon graduation. Consequently, Einstein had to start his career as a junior examiner at the Swiss patent office, a job he actually found quite fulfilling, recalling his seven years there as one of the happiest times in his life. He wrote his first important papers while employed there (including the one on special relativity and a treatise on the photoelectric effect, for which he won the Nobel Prize in 1921). But once his status as a physicist vastly improved with those publications under his belt, he was able to leave the patent office in 1909 to carry out a series of university appointments in both Zurich and Prague. He attained the peak of professional recognition in 1914, when he moved to the prestigious University of Berlin as a research professor and became a member of the Prussian Academy of

Sciences. It was over those many years of settling and resettling that he waged his mental battle with general relativity, amid teaching responsibilities, a failed marriage, and World War I. For nearly a decade, he struggled with the problem of recasting Newton's laws of gravity in the light of relativity.

He didn't go straight to the equations and fiddle with them. That wasn't his style at all. What he did at the outset was think—and think hard. Einstein knew that he first needed to establish a theoretical framework that could match what is experienced in the world around us. He carried out a variety of thought experiments in his head to see where they would lead. "Like a child who builds a toy house with blocks of various colors," explains science historian Jean Eisenstaedt, "Einstein started with sets of principles, the conceptual blocks or theoretical elements that he could place, move, suppress, and arrange in different ways; these are the bricks he used to erect his theoretical buildings."

Albert Einstein around the time that he was working on general relativity in the early 1910s. (*Hebrew University of Jerusalem, Albert Einstein Archives, courtesy of American Institute of Physics Emilio Segrè Visual Archives*)

Einstein first recognized that the force we feel upon a uniform acceleration and the force we feel when under the control of gravity are one and the same. In the jargon of physics, gravity and a constant acceleration are "equivalent." There is no difference between being pulled down on the Earth by gravity or being pulled backward in an accelerating car. To arrive at this conclusion, Einstein imagined a windowless room far out in space, magically accelerating upward, moving faster and faster with each passing second. Anyone in that room would find their feet pressed against the floor. In fact, without windows to serve as a check, you couldn't be sure you were in space. From the feel of your weight, you could as easily be standing quietly in a room on Earth. Both the magical, accelerating space elevator and the Earth, with its gravitational field keeping you in place, are equivalent systems. Einstein reasoned that the fact that the laws of physics predict exactly the same behavior for objects in the accelerating room and in Earth's gravitational hold means that gravity and acceleration are, in some fashion, the same thing.

These thought experiments, which Einstein carried out liberally to get a handle on his questions, led to some interesting insights. Watch someone throw a ball outward in that accelerating elevator in space and the ball's path will appear to you, situated outside, to curve downward as the elevator moves upward. A light beam would behave in the same way. But since acceleration and gravity have identical effects, Einstein then realized that light should also be affected by gravity, being attracted (bent) when passing a massive gravitational body, such as the Sun. The nearby matter makes a light beam's path *curve*.

Driven by his powerful physical intuition, Einstein began to pursue these ideas more earnestly around 1911. At that time he was beginning to confirm that clocks would slow down in gravitational fields. Special relativity already suggested that a *moving* clock would run more slowly; now Einstein was stating that a stationary clock would also tick more slowly when immersed in a gravitational field, an effect never before contemplated by physicists. He was saying that a clock in space will tick faster than one weighed down by Earth's gravity.

He was also coming to recognize that his final equations would likely be "non-Euclidean"—that is, based on a geometry different from the one you likely learned in grade school with basic axioms set down by the famous Greek mathematician Euclid in the third century BCE. In Euclid's world, space is entirely flat in all directions—an unchanging vista. But it was slowly dawning on Einstein that gravity would involve curvatures of space. Or to put it more correctly, curvatures of *space-time,* that invention by Minkowski that he had so blithely dismissed a few years earlier. Einstein was finally coming to appreciate Minkowski's mathematical take on special relativity and its creation of that "banal" four-dimensional manifold. Without Minkowski's earlier contribution, admitted Einstein contritely, the "general theory of relativity might have remained stuck in its diapers." Minkowski, unfortunately, was not around to hear that apology; he had died in 1909 of appendicitis at the age of forty-four.

By the summer of 1912, Einstein was at last eager to mold his burgeoning conjectures into proper mathematical form. Ignorant of non-Euclidean geometries, though, he joined up with mathematician Marcel Grossmann, an old college chum, to aid him in mastering the intricacies of this new mathematics. "Grossman," cried out Einstein upon arriving at his friend's home in Zurich, "you must help me or else I'll go crazy." Einstein chose well in seeking out assistance. It was Grossmann who pointed out to Einstein that his ideas would best be expressed in the specific geometric language first developed by the German mathematician Bernhard Riemann in the 1850s and later extended by German and Italian geometers. By 1914, Einstein moved to

Berlin and continued on his own, inexorably amending and tweaking his solutions, but now he was additionally armed with the mathematical insights Grossman had introduced him to.

His progress, though, was sluggish, and by the following year he was increasingly frustrated. His theory, as it then stood, could not accurately account for a particular shift in the orbit of Mercury. From his earliest days of contemplating a general theory of relativity, Einstein knew that a successful formulation of a new law of gravity would have to account for that anomaly.

Why is that so? Let me explain. Mercury, a planet positioned about thirty-six million miles (fifty-eight million kilometers) from the Sun, slowly revolves around the Sun, as do all the other planets. Yet these planetary orbits are not perfectly circular (as Kepler discovered) but more elliptical. With that in mind, imagine Mercury's orbit as an elongated ring. The point of this extended ring that is closest to the Sun—what is known as a planet's perihelion—shifts around over time. For Mercury the perihelion advances about 574 arcseconds each century (about 0.04 percent of its orbital circumference). Most of this tiny shift is due to Mercury's interaction with the other planets; their combined gravitational tugging alters the orbit's alignment. But that can account for only 531 arcseconds. The remaining 43 arcseconds (as measured today) were left unexplained, a nagging mystery to astronomers for decades. Newton's laws couldn't resolve the discrepancy, at least given the known makeup of the solar system. That led some to speculate that Venus might be heavier than previously thought or that Mercury had a tiny moon. The most popular solution suggested that another planet, dubbed "Vulcan" for the Roman god of fire, was orbiting closer to the Sun than Mercury, providing an extra gravitational pull. There were even a few reports of Vulcan sightings, but none were reliable.

Einstein wanted his general theory of relativity to explain that added little gravitational nudge, once and for all. As his equations stood in early 1915, Einstein was able to predict an extra shift in Mercury's orbital motion of 18 arcseconds (five thousandths of a degree) per century. But he was aiming for the measured change that was around twice as large. Discouraged, he went back and reviewed his previous work over the years. It was then that he noticed a mistake in one step of the derivations he had earlier conducted with Grossmann, a tactic the two had ultimately dismissed at the time. This spurred Einstein to reconsider that approach once again. He began to modify the equations and in the process became aware of some earlier misunderstandings. His years of toil and vexation were about to end.

His major effort took place over the course of November 1915. On each of the four Thursdays of that month he reported his incremental progress to the Prussian Academy. A breakthrough came soon after his second report on November 11. That week he was at last able to successfully calculate Mercury's extra orbital shift. He would later write a friend that he had palpitations of the heart on seeing this result: "I was beside myself with ecstasy for days." Here was the theory's first experimental success, grounding it in the real world. Moreover, Einstein's new formulation also predicted that starlight would get deflected around the Sun *twice* as much as he had earlier calculated (and twice the amount if Newton's theory is used). That's because Newton's laws take only space into account; Einstein now understood that gravity affects both space and time alike, hence doubling the effect.

Triumph arrived on November 25, the day he presented his concluding paper, titled "The Field Equations of Gravitation." In this culminating talk he presented the final modifications to his theory, one more term added in to complete the job. At last, he no longer needed a special frame of reference. He had arrived, truly and without question, at a *general* theory of gravity. The month's flurry of computational activity had been exhausting. In a letter to his longtime friend Michele Besso shortly afterward, Einstein wrote that his "boldest dreams have now been fulfilled," signing off, "your contented, but rather worn out Albert."

We usually visualize gravity as a force, something pushing or pulling on us. But Einstein introduced a new way to think of gravity—not as a force but instead as an inherent response to curvatures in space-time. From this viewpoint, objects that appear to be controlled by a force are actually just following the natural, curved pathways before them. Light, as it gets bent, is following the twists and turns of the space-time highway. And Mercury, being so close to the Sun, has more of a "dip" to contend with, which partly explains the extra shift in its orbit.

How is that so? From Einstein's perspective, space is not simply an enormous empty expanse but instead a sort of boundless rubber sheet, a physical entity unto itself. And given that image, such a sheet can be manipulated in many ways: It can be stretched or squeezed; it can be straightened or bent; it can even be indented in spots. So, massive stars like our Sun sit in this flexible mat like cosmic bowling balls, creating depressions. The more massive the object, the deeper the indent. As a consequence, planets circle the Sun, not because they are held by invisible tendrils of force, as Newton had us think, but because they are simply caught in the natural hollow carved out by the star.

It's true for smaller celestial bodies as well. Earth, for instance, is not holding onto an orbiting satellite with some phantom towline. Rather, the satellite is moving in a "straight" line—straight, that is, in its local frame of reference—the four dimensions of space-time, which are impossible for our three-dimensional minds to fully picture. But we can try with two dimensions.

General relativity says that space-time is like a vast rubber sheet. In this two-dimensional depiction, masses such as the Earth indent this flexible mat, curving space-time and generating the force we call gravity. (*Johnstone, courtesy of Wikimedia Commons*)

Think of two ancient explorers, who imagine the Earth as flat, walking directly north from the equator from separate locations. They move not one inch east or west but only push parallel to each other northward. But then they hear that they are moving closer to each other. They might conclude that some mysterious force is pushing them together. A space traveler perched high above, however, knows what's really happening. The Earth's surface is, of course, curved, and so the two explorers are merely following the spherical contour. In spherical space, two straight lines, originally parallel, *will* cross (unlike in flat space). In a similar fashion, a satellite is following the straightest route in the four-dimensional warp of space-time carved out by the Earth. As long as a heavenly body continues to exist, the indentations it creates in space-time will be part of the landscape of the cosmos. What we think of as gravity—the tendency of two objects to be drawn toward each other—can be pictured as a result of these indentations. In other words, space-time and mass-energy are the Siamese twins of the cosmos, each acting and reacting to the other. As physicist John Wheeler always liked to say, "Spacetime tells matter how to move; matter tells spacetime how to curve."

Consequently, Newton's empty box was suddenly gone. Space was no longer an inert and empty arena, as imagined since the dawn of time. Einstein showed us that space-time, this new physical quantity introduced to physics, is a real-time player in the universe at large. Thinking back on this accomplishment in his autobiographical notes, Einstein wrote, "Newton, forgive me."

No apologies needed. Einstein didn't completely overturn Newton's law of gravity. Newton got us to the Moon and back just fine, thank you, for gravity is the weakest force in our everyday life. Remember that a tiny magnet can easily pick up a paperclip against the entire pull of Earth's gravity. Newton's equations can handily deal with gravity in that environment.

What Einstein did was extend the law of gravity into realms formerly inaccessible, situations where gravity's pull is so strong—so monstrously powerful—that it causes matter to fall in at near the speed of light. Newton's laws completely break down at that point. General relativity is required when there are immense concentrations of gravitation on hand: in the world of stars, galaxies, and the universe at large, where gravity is king.

And Einstein offered an answer to the question that Newton thought unanswerable: there *is* a mechanism for gravity's effects.

Every object is just following the warps that other masses indent into space-time.

Einstein's success in accounting for that tiny, unexplained shift in Mercury's orbit was certainly a victory for his new general theory of relativity but not a fait accompli. That awaited confirmation that a light beam indeed bends by his predicted amount as it passes by a massive object like the Sun. Even as he was working on general relativity, Einstein in 1911 had suggested a specific test that astronomers could perform to confirm this response to space-time's curvatures: photograph a field of stars at night, then for comparison photograph those same stars when they pass near the Sun's limb during a solar eclipse. A beam of starlight passing right by the Sun would be gravitationally attracted to the Sun and so curve in a bit. Upon photographing that star, it would then appear to have shifted its standard position on the sky, the position it would have if the Sun wasn't in the way.

Three solar-eclipse expeditions were launched to carry out the test, but all were unsuccessful due to either bad weather or interference from the ongoing war in Europe. Data comparison problems plagued the results of a fourth test, an American effort led by astronomers from California's Lick Observatory, which were never published. That was a lucky break for Einstein. The dubious Lick results went against him, and some of the other expeditions had been carried out when his theory, still in the works, was predicting a smaller, incorrect deflection.

That's why all eyes were on British astronomers when they announced that they would try to measure the stellar shift in 1919 during a solar eclipse whose path traversed South America and crossed over to central Africa. Arthur Eddington, a renowned astrophysicist best known for his work on stellar physics, led the government-sponsored

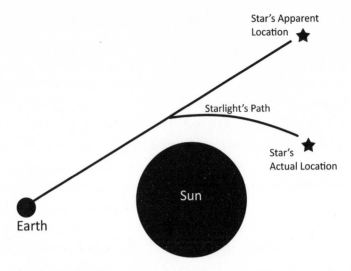

The same star field depicted when near the Sun and when away from the Sun. A star's light follows a curved path as it passes close to the Sun. But with our eyes tracing the star's light back as a straight-line path, it appears to us that the star has shifted its position in the celestial sky. (*The Cosmic Times Team, NASA Goddard Space Flight Center*)

mission to tiny Príncipe Island, off the coast of West Africa. To minimize the risk of bad weather, two other astronomers journeyed to the village of Sobral in the Amazon forests of northern Brazil. With telescopes and cameras in hand, each team was hoping to measure the extremely tiny effect. Einstein calculated that a ray of starlight just grazing the Sun's surface should get deflected by a mere 1.7 arcseconds (one thousandth the width of the Moon). Put another way, that's roughly the width of the graphite in a pencil seen from across an American football field.

On the day of the eclipse, May 29, Eddington and his assistant took sixteen photographs, most of them ultimately useless because of intervening clouds. "We have no time to snatch a glance at [the Sun]," Eddington wrote of his adventure. "We are conscious only of the weird half-light of the landscape and the hush of nature, broken by the calls of the observers, and beat of the metronome ticking out the 302 seconds of totality."

Fortunately, two of their pictures had decent images of the essential stars. For several days afterward, Eddington spent the daytime hours taking the first stab at measuring those stars. He and his companion carefully compared their pictures to another photo of the same celestial region, one taken months earlier in England at night when the Sun was far below the horizon. Eddington, an early supporter of relativity, freely admitted he was unscientifically rooting for Einstein and so was elated to see that the stars near the Sun did appear to have shifted their positions by an amount that matched Einstein's prediction, give or take several percent. It certainly didn't match the shift calculated from Newton's laws. Here was evidence that the streams of starlight were indeed bending around the darkened Sun, following its indentation in space-time. The Sobral expedition in Brazil, which had fine weather and many more photographs, confirmed Eddington's finding once they were all back in Great Britain and thoroughly examined all the images.

The results were officially announced the following November at a special joint meeting in London of the Royal Society and the Royal Astronomical Society. On the wall behind the lectern hung a picture of Isaac Newton, whose historic law of gravitation was undergoing its first big modification. The news from the meeting quickly flashed around the globe. "LIGHTS ALL ASKEW IN THE HEAVENS," blared the headline in the *New York Times*. "Men of Science More or Less Agog Over Results of Eclipse Observations . . . Stars Not Where They Seemed or Were Calculated to Be, but Nobody Need Worry."

By then Einstein was forty years old, and his life in public was never the same again. His bushy mustache, helter-skelter hair, and world-weary eyes made him instantly recognizable wherever he went. Celebrities, from presidents to movie stars, clamored to wine and dine the man whose name was now synonymous with "genius."

In a letter to Max Born in 1920, Einstein compared himself to King Midas: "Like the man in the fairy tale, whose touch turned everything into gold, thus it is with me, with everything turning into banner-line news: *suum cuique* [to each his own]." For a man of thought, who yearned for a haven of quiet to contemplate his physics, it was a state of affairs that he deemed a dazzling misery. "I really did consider flight . . . ," he continued to Born. "Now I just think about purchasing a sailboat and a little cottage near Berlin by the water."

3

One Would Then Find Oneself . . . in a Geometrical Fairyland

While the results of the 1919 solar-eclipse expedition played a large role in bringing Einstein to the public's attention, not to mention worldwide fame, there was an even earlier triumph for general relativity in the halls of academia. It had to do with the way in which general relativity's equations (a set of ten, exceedingly complex to work with) could be solved. Einstein had arrived at his first predictions based on approximations of the gravitational field around the Sun. He made some simplifying assumptions about his equations in order to make them easier to manage. Only in that way could he estimate the shift in Mercury's orbit and the amount of bending as starlight passed close by the Sun. To Einstein, an exact solution, a result that captured the entire physics and mathematics of the problem without involving approximations, appeared insurmountable. But to his surprise, that wasn't the case at all.

Very soon after Einstein's final presentation before the Berlin Academy—in less than a month, in fact—the German astronomer Karl Schwarzschild arrived at the first *full* solution to general relativity's equations. He sent his findings to Einstein immediately, believing that he was allowing "Mr. Einstein's result to shine with increased purity," as he noted in his report. It was this remarkable achievement, which both surprised and delighted Einstein, that initiated the long march toward our modern conception of the black hole.

Both a practical astronomer and a theorist, Schwarzschild was a stand-out in a multitude of fields. He made major contributions in electrodynamics, optics, quantum theory, and stellar astronomy; he was a pioneer in substituting photographic plates for the human eye at the telescope; and he could at times be quite bold in his speculations. Fifteen years before Einstein even introduced the notion of space-time bending, Schwarzschild had pondered whether space was curved rather than flat—either turned inward like a sphere or curved outward like a hyperbola out to infinity. "We can wonder how the world would appear in a spherical or a pseudo-spherical geometry . . . ," he told a meeting of German astronomers in 1900. "One would then find oneself, if one will, in a geometrical fairyland; and one does not know whether the beauty of this fairyland may not in fact be realized in nature." No wonder Schwarzschild latched onto Einstein's equations so quickly; he had been anticipating them (and avidly followed Einstein's progress in developing general relativity) over the years.

As director of the Potsdam Astrophysical Observatory, then the most esteemed position a German astronomer could attain, Schwarzschild wanted to remove all doubt as to the uniqueness of Einstein's results. And in aiming for that goal, he ended up devising a method that became a valuable tool for relativists for years afterward.

In carrying out this endeavor, Schwarzschild did what all good mathematicians do—devise a scheme that makes the mathematics of the problem simpler. For one, he used spherical coordinates, which makes it easier to map the gravitational field around a spherical mass—in this case, a nonspinning star. To see how this approach can make a complex question simpler, imagine an everyday problem:

Karl Schwarzschild
(*American Institute of Physics Emilio Segrè Archives, courtesy of Martin Schwarzschild*)

Take an airplane circling an airport from three miles (almost five kilometers) away. If you want to describe its path using the geometry of a flat grid, the result is very messy. If you designate its east-west position x and its north-south position y, then the algebraic equation that describes its entire route is $x^2 + y^2 = 3^2$. But let's say you shift to a different geometry altogether: a graph with radial, or circular, coordinates. In that case you don't have to worry about x's and y's at all. The plane is always three miles from the center of a circle, and the equation that describes its flight path is no more complicated than $r = 3$ (radius = 3). That, in a way, is what Schwarzschild did.

But his new set of coordinates led to a whopping predicament when he looked at the very center of space-time, where his star resided. As Ralph Sampson, the astronomer royal for Scotland, remarked at the time, "The consequences . . . are so startling that it is difficult to believe they have any relationship to reality." To understand this dilemma, picture what happens if all the mass of that star, say the Sun, is squeezed down to a very small size. Upon doing this, Schwarzschild discovered that, around this hypothetical point, a spherical region of space suddenly arose out of which nothing—no signal, not a glimmer of light nor bit of matter—could escape. In its day, it was called "Schwarzschild's sphere." Today we call this boundary the "event horizon." That's because no event occurring within its borders can be observed from the outside. More than an indentation, space-time in this case becomes a bottomless pit. Light and matter can go in but never come back out. It's a point of no return. The light and matter get crushed down to a singular point, a condition of zero volume and infinite density called a "singularity." It's where the ordinary laws of physics completely break down.

But I'm getting ahead of myself. That is how we currently picture such a singularity. Schwarzschild and others in his day actually viewed this situation fairly differently: watching how objects, such as light particles, approached Schwarzschild's sphere, "they got stuck, so to speak," explains historian Eisenstaedt. "This was then taken as bona fide evidence that all trajectories ended up or died at the [sphere], where time . . . stopped. . . . The [light's] trajectory appeared to perpetually approach the magical sphere, as if to vanish there." Or maybe they simply piled up on the surface of this magical ball. It was a strange and weird place. The "Schwarzschild singularity" (as it was also called) was an impenetrable sphere from their perspective.

Arthur Eddington, in his 1926 book *The Internal Constitution of the Stars,* was confident that no star could possibly collapse to such a compacted state. So, why worry about it? As he fancifully put it, "The mass would produce so much curvature of the space-time metric that space would close up round the star, leaving us outside (i.e. nowhere)."

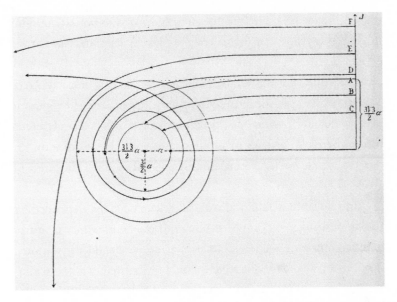

A 1924 illustration of various light beams approaching a Schwarzschild singularity (empty sphere in the center). Those that couldn't escape just vanished at the surface where time stopped. (*From Max von Laue,* Die Relativitätstheorie, *volume 2, 1924*)

That was one way to look at it. But despite Eddington's imaginative description, most relativists at the time did not seriously think that space-time itself was being significantly warped and twisted around Schwarzschild's singularity. "They realized that the spatial component might be slightly bent, that time may be a bit out of step, but nobody imagined that Schwarzschild's solution could represent a space *really* different, completely different, from Newton's," explains Eisenstaedt. That awaited new mathematical insights in the 1960s. Relativists needed the ability to map the entire region of space-time around the singularity—a grand, calculation-intensive enterprise impossible for physicists in the 1910s and 1920s to undertake. The modern-day vision of a "black hole," a pit in space-time, was not yet imagined.

But, still, what was the best way to describe this unusual place? Schwarzschild used the term *discontinuity*. In France and Belgium it became the *sphère catastrophique,* for it did appear like a catastrophic place where all the laws of physics went awry. For Eddington, it was the "magic circle." Others simply referred to it as a "frontier" or "barrier."

And how big would that magical sphere be? That depended on the amount of mass caught inside it. If our Sun, which is nearly 900,000 miles (1.4 million kilometers) wide, were suddenly squeezed down to a point, its magic sphere would be less than 4 miles (6 kilometers) across. But the Earth, perched some 93 million miles (150 million kilometers) away, would not be affected at all. Indeed, all the planets would still orbit around the Sun in the same way they have for some four billion years if the Sun were that size. Although the Sun's mass is more condensed, it exerts the same gravitational pull on us. It is only closer in that the gravitational pull of the magical sphere starts to soar.

What happens if the amount of mass is greater, such as the equivalent of ten suns squished down to a point? In that case, the magic sphere would stretch almost forty miles (around sixty kilometers) across. The equations showed that the width of the magical sphere (that is, the event horizon) expands as more and more mass is trapped within it.

Einstein didn't expend a lot of energy worrying about these singularities. He figured that Schwarzschild's novel entity was just a sign that general relativity was still incomplete, that such a hazardous outcome would disappear after he had formulated a unified theory of gravity and electromagnetism, a venture on which he spent the rest of his life with no success.

Many viewed Schwarzschild's sphere as merely an artifact, one that arose because of the coordinates being used but had no physical significance whatsoever. Others were not worried for practical reasons. Why be concerned about all that mass being squeezed below its event horizon, when no star was ever seen to be that small? This never happens, they said. Nature was surely providing the means to save the day. Schwarzschild himself figured that the pressure pushing back from all that squeezed matter would step in to prevent the collapse in the first place. Einstein thought so, too. He even worked out a little calculation at a Paris meeting in 1922 showing how a star's pressure would prevent a large star from catastrophically collapsing. Moreover, no one at this time thought that densities could ever be greater than when atoms are packed as tightly as possible. But Schwarzschild wasn't even predicting the existence of such a situation; to him, his clever theoretical setup simply enabled him to obtain an exact solution for Einstein's equations of general relativity and more easily map the gravitational field around a star. It was all just a mathematical game. As he reported, the problem of infinite pressures at the center was "clearly not physically meaningful."

Schwarzschild's accomplishment was even more astounding when you consider the circumstances in which he carried out his calculations. It was at the height of World War I, and Schwarzschild, serving as a lieutenant in the German army, was then posted on the Russian front. His job was to calculate trajectories for long-range projectiles. Relativity was on his mind because, while on leave, he had been in the audience at the Prussian Academy on 18 November 1915 when Einstein presented his successful calculation of Mercury's perihelion advance. Previously cautious that Einstein's ideas could be astronomically verified, he was now convinced. On returning to the battlefront, he received copies of Einstein's finalized theory and swiftly completed two papers on the subject. As he wrote in his letter to Einstein, "As you see, the war is kindly disposed toward me, allowing me, despite fierce gunfire at a decidedly terrestrial distance, to take this walk into this your land of ideas." Known for his warmth and outgoing personality (always ready for a good beer to quench his thirst), Schwarzschild seems to have retained this demeanor even in the midst of war.

Schwarzschild's first battlefront solution was communicated to the Berlin Academy on 13 January 1916 by Einstein himself, who described the result as splendid. "I would not have expected that the exact solution to the problem could be formulated so simply," replied Einstein to Schwarzschild. "The mathematical treatment of the subject appeals to me exceedingly."

Sadly, Schwarzschild didn't have long to bask in Einstein's praise. While in the trenches, he contracted pemphigus, a rare and then fatal autoimmune disease that attacks the skin. Critically ill, the noted astronomer returned to Potsdam in March 1916 and succumbed to his illness on 11 May, just four months after his academy triumph. He was forty-two years old.

Though Schwarzschild may not have thought that his solution applied to the real universe, others considered the possibility. Writing in the *Philosophical Magazine* in 1920, the Irish physicist Alexander Anderson of University College, Galway, pondered what would happen if the Sun's girth were to contract to below its magical sphere width. This was a time when many still thought that the Sun generated its energy through slow gravitational contraction. So, should the Sun keep on shrinking, wrote Anderson, "there will come a time when it will be shrouded in darkness, not because it has no light to emit, but because its gravitational field will become impermeable to light."

Contemplating such a gravitational collapse was a prescient thought, yet few followed up on it. One notable exception was the British physicist Sir Oliver Lodge, who in 1921 also noted that the gravity of a sufficiently dense star would prevent its light from escaping. This would happen, he noted, if the mass of the Sun were squeezed into a globe some three kilometers (just under two miles) in radius. "But," he concluded, "concentration to that extent is beyond the range of rational attention."

However, although Lodge doubted that a single stellar nugget would form on its own, he boldly imagined that the light-swallowing effect might occur with larger collections of celestial matter. "A stellar system—say a super spiral nebula—of aggregate mass equal to 10^{16} suns . . . might have a group radius of 300 parsecs [around 1,000 light-years] . . . without much light being able to escape from it. This really does not seem an utterly impossible concentration of matter." He was right. Lodge was crudely anticipating the existence of supermassive black holes, the kind found in the centers of most galaxies.

But all these speculations went nowhere and wouldn't be reconsidered for another two decades. The concept of the black hole—or, more correctly, its early equivalent—was still in its infancy. What pushed it along was the discovery of strange, new stars in the celestial sky—of a type that no astronomer had ever anticipated beforehand.

4

There Should Be a Law of Nature to Prevent a Star from Behaving in This Absurd Way!

When Schwarzschild singularities were first introduced, they were essentially theoretical oddities. No one really expected them to jump off the page of a scientific journal into the real world. But new and startling astronomical findings in the early twentieth century made it difficult for theorists to maintain that attitude for long. If any single observation could be blamed, it would be the one concerning the faint companion that slowly circles Sirius, the brightest star in the nighttime sky and long known as the "Dog Star" because it's situated in the northern constellation Canis Major.

The story of Sirius and its tagalong commences in the nineteenth century at the Königsberg Observatory in Prussia, where Friedrich Wilhelm Bessel was setting new standards in positional astronomy. The observatory director had already gained fame in 1838 for being the first person to directly measure the distance to a star, astronomy's biggest challenge at the time. Afterward, he turned his attention to stellar movements.

For a number of years Bessel went through old stellar catalogs, as well as making his own measurements, to track how the stars Sirius and Procyon were moving over time across the celestial sky. By 1844 he had enough data to announce that Sirius and Procyon weren't traveling smoothly, as expected; instead, each star displayed a slight but distinct wobble—up and down, up and down. With great cleverness, Bessel deduced that each star's quivering must result from the pull of a dark, invisible object circling it, like a little boy tugging on his mother's skirt. Sirius's companion, he estimated, completed one orbit around Sirius every fifty years.

Bessel was clearly excited by his find; in his communication to Great Britain's Royal Astronomical Society he wrote, "The subject . . . seems to me so important for the whole of practical astronomy, that I think it worthy of having your attention directed to it."

Astronomers did take notice, and some tried to discern Sirius's companion through their telescopes. Unfortunately, at the time Bessel reported his discovery, Sirius B (as the tiny companion came to be known) was at its closest to gleaming Sirius (from the point of view of an observer on Earth) and thus lost in the glare. But even years later, no one was successful in spotting the bright star's partner.

That all changed on 31 January 1862. That night in Cambridgeport, Massachusetts, Alvan Clark, the best telescope manufacturer in the United States, and his younger son, Alvan Graham Clark, were testing the optics for a new refractor they had been building for the University of Mississippi. It was going to be the biggest refracting telescope in the world. Looking at notable stars to carry out a color test of their 18.5-inch lens, the son observed a faint star very close to Sirius.

This momentous sighting might have gone unrecorded. But fortunately the father was an avid double-star observer and possibly encouraged his son to report the discovery to the nearby Harvard College Observatory. In fact, according to historian Barbara Welther, rather than its being an accidental discovery, as long asserted in astronomy books, "there might have been a [prearranged] connection between the elder Clark and someone at Harvard" to look for Sirius's companion.

Whatever the case, George Bond, the observatory's director, confirmed the find a week later, and he soon wrote up two papers: first a brief notice to a German journal of astronomy, then a more candid report to the *American Journal of Science*. This second paper revealed that one question was uppermost on Bond's mind: "It remains to be seen," he wrote, "whether this will prove to be the hitherto invisible body disturbing the motions of Sirius." The newfound star seemed to be in the right place to explain the direction of Sirius's wavelike motions, but its luminosity was extremely feeble—so dim, in fact, that it suggested at the time a star too small to have enough mass to account for the wobble. Here was the first clue to Sirius B's uniqueness.

For revealing Sirius's dark companion, Alvan Graham Clark in 1862 garnered the prestigious Lalande Prize, given by the French Academy of Sciences for the year's most outstanding achievement. Astronomers around the globe continued over the years to observe the orbital dance of Sirius and its partner and eventually determined that the companion was hefty enough (an entire solar mass) to pull on Sirius, though with a light output less than a hundredth of our Sun's. But no one worried about this disparity right away. They just shrugged their shoulders and figured that Sirius B was a sunlike star cooling off at the end of its life.

At this point, no one had yet secured a spectrum of the light emanating from Sirius B—in other words, a breakdown of the distinct wavelengths emanating from the tiny orb. This was a difficult task owing to the overwhelming brightness of the binary's primary star. Until they could obtain the star's spectrum, astronomers just assumed that Sirius B was either yellow or red, like other dim and cooler stars.

That's because astronomy had a general rule at the time: the hotter the star, the brighter. The brightest stars' colors were white, blue-white, or blue.

But in 1910, Princeton astronomer Henry Norris Russell noticed something that cast doubt on that rule. On a Harvard Observatory photographic plate, a faint companion of the star 40 Eridani—a companion known since 1783—was labeled as blue-white. Russell doubted that such a classification could be correct, but in 1914, Walter Adams at the Mount Wilson Observatory in California confirmed the spectrum. How could a star be white-hot, yet dim? "I was flabbergasted," recalled Russell. "I was really baffled trying to make out what it meant." Then, in 1915, Adams determined that Sirius's faint companion, too, displayed the spectral features of a blazing blue-white star—at 25,000 Kelvin, far hotter than our Sun. So why wasn't it as fiercely bright to our eyes as Sirius itself? How could a fiery, white star emit such paltry radiation? If such a star replaced the Sun, it would appear only one four-hundredth as bright to us.

Arthur Eddington (*American Institute of Physics Emilio Segrè Visual Archives, gift of Subrahmanyan Chandrasekhar*)

Soon theorists, such as the Estonian Ernst Öpik and the British astrophysicist Arthur Eddington, figured out what was going on. If a star is both white and hotter than our Sun, it must be emitting more light over each square centimeter of its surface. But since Sirius B is so faint, that could only mean it had less surface area than our Sun—in other words, it is far denser and smaller. It fact, it had to be just a little larger than the size of the Earth. (So astounding was the density he calculated, around twenty-five thousand times more than the Sun, that Öpik at first declared it "an impossible result.") Such stars came to be called "white dwarfs."

With a Sun's worth of mass being squeezed into such a tiny volume, astronomers and physicists alike were powerless to explain how a star could remain stable in this incredibly compressed state. Physics at that stage couldn't explain how such densities could be continually sustained. As Eddington later remarked mischievously, "The message of the Companion of Sirius when it was decoded ran: 'I am composed of material 3,000 times denser than anything you have ever come across; a ton of my material would be a little nugget that you could put in a matchbox.' What reply can one make to such a message? The reply which most of us made . . . was—'Shut up. Don't talk nonsense.'"

It took quantum mechanics, under development in the 1920s, finally to solve the puzzle. With the mass of the entire Sun crushed into an Earth-sized space—creating the densest matter then known in the universe—British theorist Ralph Fowler in 1926 figured out that pressures inside the compact dwarf star become so extreme that all its atomic nuclei, like droves of little marbles, are packed into the smallest volume possible. Atoms are largely empty space. (If an atom were blown up to the size of a football stadium, the nucleus would look like a pea perched on the fifty-yard line, with the tiny electrons buzzing around the farthest seats.) But all that extra space is drastically reduced within a white dwarf star. At the same time, its free electrons generate an internal energy and pressure that keep the atom from collapsing further. With the electrons elbowing one another with a vengeance (a quantum mechanical rule formulated by Wolfgang Pauli forbids them to merge), they resist additional compression. And that is the key to a white dwarf's stability: the incredible pressure exerted by the highly confined and fast-moving electrons—known as a "degeneracy pressure"—prevents the star from further compaction. This pressure exceeds the crushing forces found at the center of our Sun a million times over. Such a pressure was inconceivable until the arrival of quantum mechanics.

A white dwarf's ultradense material is impossible to assemble here on Earth; only the star's extreme environment makes it feasible. Astronomers later learned that such densities are always the end stage for a star like our Sun. The white dwarf is the luminous stellar core left behind after the star runs out of fuel and releases its gaseous outer envelope into space. Such will be our Sun's fate some five billion years from now. Radiating the energy left over from its fiery past, the white dwarf, like a dying ember, eventually cools down and fades away.

The discovery of the extremely dense white dwarf star turned out to be only the first volley in a startling stellar revolution. By the 1930s, working with the new laws of both quantum mechanics and relativity, theorists were astonished (and disturbed) to find that dying stars, if they had enough mass, might face even stranger fates than turning into a white dwarf. The discovery of the white dwarf—and the understanding of its physics—opened up a whole can of cosmic worms.

The first steps in opening up that can were taken in the summer of 1930, just as the global economy was sinking into its Great Depression. The advance commenced during an eighteen-day sea voyage from India. The traveler was nineteen-year-old Subrahmanyan

Chandrasekhar, a dignified youth known by one and all as simply Chandra. While journeying to England, first by ship and then by train, to begin his graduate studies with Ralph Fowler at Cambridge University on scholarship, he explored the physics of white dwarf stars, a subject he had become enchanted with during his earlier studies at the University of Madras. Before his departure he had already prepared a paper on the density of white dwarfs, which combined Fowler's idea with Arthur Eddington's model of a star.

Fowler had just shown how the pressure from electrons, tightly packed in the compact star at a density of a metric ton per cubic centimeter, keeps a white dwarf intact. But can this go on forever? Chandra further mused on the ship. What happens, asked the young student, if a white dwarf is more massive? With plenty of time to think on the long voyage, which took him through the Suez Canal into the Mediterranean, Chandra had an epiphany. He came to realize that as the white dwarf got heavier and heavier, many of the electrons moving within the dense stellar nugget would approach the speed of light. And that meant it was necessary to apply the rules of relativity to the star's behavior, something Fowler did not do.

Having acquainted himself with quantum mechanics and relativity as an undergraduate, Chandra carried out a calculation right there on the steamer and, to his surprise, concluded that there is a maximum limit to the mass of a white dwarf (now known to be 1.4 solar masses). An avid reader of the scientific literature, he happened to have read the key books that allowed him to work this out and had three with him onboard to consult. "It is something which is so simple and elementary that anyone could do it," Chandra modestly recalled in 1971. Past his calculated limit, the white dwarf star could not support itself against gravity. What exactly occurred at that boundary was completely unknown territory. Chandra had no idea what the white

Subrahmanyan Chandrasekhar at Cambridge University in 1934. (*American Institute of Physics Emilio Segrè Visual Archives, gift of Subrahmanyan Chandrasekhar*)

dwarf would turn into if it were heavier. "I didn't know how it would end," he said further, thinking back on that moment of discovery. Yet he quickly wrote a paper on the finding upon his arrival.

Fowler communicated Chandra's pre-voyage paper on the density of white dwarfs to the *Philosophical Magazine* but sent his relativistic solution to another expert to assess. After waiting months with no feedback, the young researcher, by then disappointed that publication in Great Britain was unlikely, went ahead on his own and mailed that second article off to America. The result was a brief 1931 paper published in the *Astrophysical Journal* and titled "The Maximum Mass of Ideal White Dwarfs," which almost got rejected. A referee had initially doubted one of Chandra's equations, until provided a detailed proof. "I am sorry that I was in error in criticizing his equation," the referee told the editor, "but it seems to me a rather remarkable thing that this equation is true. I should not have expected it at the first glance."

When Chandra first started on his calculations, he didn't know that others—Edmund Stoner in England and Wilhelm Anderson in Estonia—had already published earlier estimates of an upper bound to a white dwarf's density. They thought of it as the tightest space that atoms could feasibly jam together. But Chandra used a more sophisticated model of a star, which in the end arrived at a stronger (and stranger) conclusion. His equations were telling him that past his threshold, the star appeared headed toward total collapse, with its density going to infinity (a result he deemed "inconceivable").

It's not surprising that Chandra was not alone in his pursuit. The problem of stellar mass was in the air. It was a time when astrophysicists were starting to analyze the internal structure of a star—how it was powered, how it was built. For centuries, astronomers had solely tracked the positions and movements of the stars; now, they wanted to crack open the star (in theory) and find out how it worked. Just as Chandra was musing on white dwarfs in England, the brilliant theorist Lev Landau was doing the same in the Soviet Union. By pondering the question of a star's inner structure, Landau thought he might make some startling new discoveries in nuclear physics, his specialty. After setting up a simplified model of a star as a lump of cold matter, he also concluded in 1931 that "there exists in the whole quantum theory no cause preventing the system from collapsing to a point," if the star were heavier than 1.5 solar masses. But this was obviously a "ridiculous" result, he decided. He knew there were stars more massive. What could possibly explain this obvious contradiction? To answer that, Landau reasoned that the laws of physics must be breaking down within the heart of a star, following up on a thought that the Danish atomic physicist Niels Bohr had earlier expressed. The stellar core was a "pathological" region, as Landau put it, where matter becomes so dense it forms "one gigantic nucleus." While seemingly prescient, it was more a harbinger of revelations to come.

Meanwhile, Chandra continued to pursue the mystery of a white dwarf star's fate, which only increased his perplexity. He published these concerns in 1932, this time in a German journal to avoid possible rejection by British referees. In the article's final sentence, he wrote, "We may conclude that great progress in the analysis of stellar structure is not possible before we can answer the following fundamental question: *Given an enclosure containing electrons and atomic nuclei (total charge zero), what happens if we go on compressing the material indefinitely?*" Indeed, what does happen to the star? By putting his final thought in italics, Chandra was likely pressing the astrophysical community, which was largely uninterested at this stage, to pay attention. The top Britishers in stellar physics—Eddington, James Jeans, and Edward Milne—were too busy arguing with one another at meetings on the exact construction and composition of a star's innards to pay attention to the theories of a lowly graduate student. The only thing the three combatants could agree on was that a star would never collapse to a point.

After the publication of his 1932 paper, travel and other astrophysical questions diverted Chandra for a while, but once he obtained his PhD and was elected to a fellowship in Cambridge's Trinity College, Chandra came back to the problem. "It is necessary to emphasize one major result of the whole investigation," he wrote in 1934, "namely, that it must be taken as well established that the life-history of a star of small mass must be essentially different from the life-history of a star of large mass. For a star of small mass the natural white-dwarf stage is an initial step towards complete extinction. A star of large mass . . . cannot pass into the white-dwarf stage, and one is left speculating on other possibilities." In other words, low-mass stars would assuredly die as white dwarfs, but what fate befalls a star of higher mass, whose central core steps over the limit? What could possibly happen to it?

The idea occurred to him that his finding might lead to new physics, "but I kept away from it . . . ," he said. "I was not willing to draw that conclusion." As a foreigner at a university with so many notable figures in physics, he often felt that he didn't belong. "It seemed to me that there were . . . far too many people doing important things, and what I was doing was insignificant in comparison. I suppose I was afraid," he recalled.

Yet Chandra quietly continued working on the problem. During a visit to Russia, Soviet scientists convinced him that no astronomer would take his limit seriously until he demonstrated that a good sample of white dwarf stars, across a range of densities and properties, never venture beyond the critical mass limit. He decided to take on the challenge, which involved solving a complex differential equation for each star using a clunky desktop calculator. In the end, Chandra completed an eighteen-page paper, teeming with calculations. Finished on 1 January 1935, it was headed for publication in the *Monthly Notices of the Royal Astronomical Society*. A graph he included in the paper visually portrayed his startling bottom-line: as a white dwarf star grew smaller and smaller with increased mass, its radius approached zero. Past a certain weight, a white dwarf star simply shriveled up toward nothingness. Chandra's earlier work had been based on approximations. This time he had an exact solution.

Achieving that goal was a draining experience for him. It involved several months of twelve-hour days. "Involved in the puzzles of the interior of stars, battered by differential equations, boxed by numerical calculations, impeded by ignorance, rushed on by the dawning of the New Year," he wrote his brother Balakrishnan soon after, "I have at last emerged not indeed with the anticipated joy with which I began [diving] into the Crucibles of Nature, but burnt and smoking, dissatisfied, tired."

His conclusion was glaringly blunt. "When the central density is high enough . . . ," wrote Chandra, "the configurations then would have such small radii they would cease to have any practical importance in astrophysics." Stars were not expected to act like this. Sir Arthur Eddington was not pleased at all by this news and, during a discussion of Chandrasekhar's idea of drastic stellar collapse at the 11 January 1935 meeting of the Royal Astronomical Society in London, made the infamous declaration (so often quoted) that "there should be a law of Nature to prevent a star from behaving in this absurd way!" The audience howled in laughter.

Having just made his presentation to the society, which had been greeted with polite applause, Chandra was horrified to hear this cutting response from Eddington and was mortified by the audience's response. Chandra had been consulting Eddington for weeks as he carried out his computations, and the great man hadn't mentioned one word of disapproval. He had even helped Chandra get his needed calculator. It appears that Eddington ignobly waited for a public forum to pounce on Chandra's results, turning it into one of the most notorious intellectual duels in the history of astrophysics.

Eddington had one overwhelming gripe: he thought it was wrong to mix special relativity with quantum mechanics, at least in the way that Chandra handled it for white dwarf stars. "I do not know whether I shall escape from this meeting alive," Eddington told the astronomers, "but the point . . . is that there is no such thing as relativistic degeneracy. . . . I do not regard the offspring of such a union as born in lawful wedlock." In a strident article in the *Monthly Notices of the Royal Astronomical Society* later that same year, Eddington continued to call it "an unholy alliance." He simply didn't trust Chandra's particular approach. Eddington had displayed such brilliance in the field of astrophysics for so long, especially in setting up the standard model of a star (one of twentieth-century astronomy's greatest accomplishments), that he couldn't imagine being wrong. With his academic tweeds, upright posture, and a prim pince-nez clipped on his nose, this noted astrophysicist appeared the very personification of that British pride.

Eddington's forthright response at the astronomical meeting was not unusual, though. He was always ready for a good intellectual rumble from time to time. To him, it was the way science got done. Chandra was not alone in his distress; many others had been slashed by Eddington's rapier wit over the years. But why didn't others at the society that fateful evening come to Chandra's defense? Part of the problem was the difficulty of the math and physics in Chandrasekhar's work; few were as familiar with stellar theories (not to mention quantum mechanics or special relativity) as Chandra was to back him up. Eddington was *the* world's expert on the structure and luminosity of stars. Surely he must be right, these bystanders assumed, and Chandra wrong. Others, while supportive of the young theorist's work, were fearful of openly criticizing Eddington, the most heralded astrophysicist of his era. Even when key astrophysicists within a few years came to recognize Eddington's error, admitting this to Chandra in private, these scientists continued to keep quiet in public, not wishing to disgrace the high priest of astronomy. Many counseled Chandra to keep any rebuttals to himself. Chandra was very bitter over this lack of support from his peers.

Given Eddington's renowned expertise on stars, it's puzzling that he was not leading the way to new astrophysics. He was a world expert on relativity and comfortable with the application of quantum mechanics in other arenas. In fact, he had earlier supported some of Stoner's findings on the maximum limit of white dwarfs, communicating them to the *Monthly Notices of the Royal Astronomical Society*. Why was Eddington now so vehement in keeping relativity and quantum mechanics apart in this case? It's possible that he was simply overwhelmed by the psychological factor—the preposterous notion that matter could somehow be crushed to oblivion. Where was all that matter going? By then fifty-two years of age and educated during an era when the known universe had been far simpler, Eddington was absolutely certain that nature didn't behave that way. For him it defied common sense. He acted as if he could smash Chandra's conclusion through a sheer act of will, ignoring any physics not to his liking. According to British science historian Arthur Miller, Eddington was mainly bullying Chandra to protect a fanciful mathematical scheme he had been working on for eight years, a cherished (and ultimately foolish) project whose aim was to naturally derive both the physical constants of nature and the number of particles in the universe. Chandra's finding put all his hard-won work into jeopardy. Eddington's unified theory didn't work if relativistic degeneracy were true.

So, it's not surprising that Eddington remained adamant in his opposition. In a 1936 address at Harvard University, he continued to call Chandra's white dwarf limit "stellar buffoonery." Chandra, ever the gentleman, faced the censure with stoic grace. In those days, says Canadian physicist Werner Israel, "debates were a sport, like cricket. Afterwards, you repaired to the Common Room and shared a glass of port." Agreeing to disagree, the two managed to maintain cordial relations, continuing to take tea, attend sports events, and go on bike rides together. Certain his analysis was correct, Chandra figured time would settle the issue in his favor. So, he was excruciatingly patient, though inwardly dismayed at what astronomers were thinking of him at the time. "They considered me as a sort of Don Quixote trying to kill Eddington," he said some forty years later. "As you can imagine, it was a very discouraging experience for me—to find myself in a controversy with the leading figure of astronomy."

To be ridiculed by one of England's towering figures was a scientific humiliation and setback for the young investigator, and it took more than twenty years before the "Chandrasekhar limit," the highest possible mass of a white dwarf star, became a fundamental parameter in astrophysics textbooks. A Nobel Prize for Chandra followed (much later) in 1983.

There was a downside to all the machinations that transpired in the 1930s: with Chandra's confidence decidedly shaken, he abandoned the topic for a couple of decades, eventually immigrating to America, where scientists were more receptive to his ideas. There, at the Yerkes Observatory and the University of Chicago, he proceeded to work on other astrophysical problems. "I had to make up my mind as to what to do. Should I go on the rest of my life fighting?" Chandra later recalled. "After all I was in my middle twenties at that time. I foresaw for myself some thirty to forty years of scientific work, and I simply did not think it was productive to constantly harp on something which was done." Though Chandra displayed a public equanimity regarding this tempest, privately Eddington's criticism stung sharply.

Eddington, it turned out, was simply wrong. Nature did not provide a safety net against stellar collapse. The young Chandra never ventured to guess exactly what would happen to a white dwarf star that ventured past 1.4 solar masses. Conservative by nature, he was never keen on speculation. But he did open the door wide for other theorists to contemplate the existence of neutron stars and black holes.

Meanwhile, we can mull over the "what if" in this tale: If Eddington had been Chandra's champion rather than foil, would astronomers have accepted the possibility of black holes faster? Not likely, says Israel, who has deeply examined the scientific temper of these times. "In 1935," he writes, "the astronomical community was not yet ready to 'buy' the idea of gravitational collapse, not even if a master salesman like Eddington had been ready to exert all of his persuasion." In this prewar era, astronomers were fairly old-fashioned. Few were trained or even interested in applying the new physics—relativity and quantum mechanics—to astrophysical problems. Many didn't think relativity was even a part of physics but more a branch of mathematics.

If a new law of physics didn't stop a singularity from forming, astronomers at this point were confident that other forces would come into play. Stellar physics was still a relatively young field with lots of unknowns. Many just figured that massive stars experienced a vast weight reduction, ejecting enough of their matter over time that each and every one ultimately fell below the critical 1.4-solar-mass limit, where they could safely die as white dwarf stars. Even Chandra admitted he leaned toward this view for a time.

But this was simply a nice "just-so" story—a convenient dodge—that only for a time kept astronomers from having to face the unimaginable.

5

I'll Show Those Bastards

Throughout the Milky Way galaxy, there are multiple star systems where two or more stars orbit one another, just the way the Moon circles the Earth. And if one of those stars happens to be a white dwarf, interesting things can happen—as on the night of 29 August 1975.

As twilight settled over Japan that day, high-school senior Kentaro Osada, an amateur astronomer who regularly scanned the heavens in his free time, noticed that the northern constellation Cygnus the Swan had a new star in its tail, a pinpoint of light that hadn't been there before and whose luminosity would soon rival the constellation's brightest star, Deneb. Within hours, a slew of other amateur and professional astronomers wired or phoned news of the appearance to the Central Bureau for Astronomical Telegrams in Cambridge, Massachusetts, the official clearinghouse for new celestial sightings.

What they were all witnessing was a nova, the name given to the event by ancient astronomers in their mistaken belief that the heavens had created a brand-new star. The appearance of a nova in the sky more than two thousand years ago inspired the Greek astronomer Hipparchus to prepare the first serious catalog of the stars as seen from the Western world. By the mid-nineteenth century, there were a number of quaint theories about a nova's origin, including swarms of meteors colliding with one another or even a star encountering a cloud of cosmic material and heated to superluminal brightness by the friction as it passed through. But what Osada and others were actually witnessing that night was a stellar outburst—a star that suddenly erupts into startling brilliance then slowly fades back to its normal brightness.

V1500 Cygni, as the 1975 nova is now officially labeled, involved a white dwarf star and its close companion, a small red star—a binary system located roughly six thousand light-years away. Because the two were such intimate neighbors, the intense gravitational field of the white dwarf was able to pull gas away from the red star, which then formed a swirling disk of matter around the dwarf. Over time, some of this material reached the surface of the dwarf, wrapping the entire orb in a thin blanket of hydrogen. Compressed and heated by gravity, the layer suddenly ignited, engulfing the white dwarf in a monstrous thermonuclear blast. The nova was born. V1500 Cygni's luminosity swiftly increased by a factor of one hundred million, making it one of the brightest novae in the twentieth century, seen for days.

Yet, despite such a violent explosion, the system remains intact. Nova Cygni might reappear in ten thousand or more years, after the system's thief, the white dwarf, steals another layer of fusible hydrogen from its mate. That is the most common form of nova in the heavens. Each year about thirty white dwarfs, sprinkled throughout our galaxy, blow off a little "steam" in this manner. "There is something uncanny about the change in the ancient and familiar configurations of a constellation," said Henry Norris Russell in 1939. "I well recall the impression which my first glimpse of a Nova produced on me. . . . It was hard to escape the feeling that it was making a noise!" Many other novae are regularly sighted each year in galaxies beyond our own.

But even before astronomers understood exactly what a nova was, they recognized that there was more than one kind. Along with those "common" novae popping off in the heavens, astronomers began noticing others that were in a class by themselves—novae that were far more luminous and much rarer. Russell floridly described them as a "phenomena of a different order—the most tremendous yet known to the mind of man." In our own galaxy, they are seen only once every few centuries. The famous Crab Nebula in the constellation Taurus is the remnant of one such burst, an event recorded by Chinese astronomers in the year 1054. Some astronomers called them "giant novae," others "exceptional novae." In his native German, Walter Baade at the Mount Wilson Observatory in California referred to them as "Hauptnovae" (chief novae).

Not as publicly famous as his Mount Wilson colleague Edwin Hubble, who had confirmed the expansion of the universe, Baade was actually the superior observational astronomer. In 1952 he corrected

Walter Baade (*Courtesy of the Huntington Library, San Marino, California*)

Hubble's too-small size of the universe, discovering that it was twice as big and twice as old. Born and educated in Germany, Baade had a hip defect that made him walk with a limp. But this disability allowed the aspiring astronomer to stay out of World War I and focus on learning a host of useful observing techniques during his graduate studies. Baade, according to his colleagues, "saw the mysteries of the universe as the greatest of all detective stories in which he was one of the principal sleuths."

Obtaining a job at the Hamburg Observatory after earning his doctoral degree, Baade got his first look at one of those superbright, rare novae in 1921, the sudden flare-up occurring in a little spiral galaxy called NGC 2608. He was immediately hooked and regularly photographed the nova until it faded away the following year. It was Baade who confirmed that such exceptional novae were stupendously more energetic than the more common type. Indeed, one of these bursts can be as bright as the combined light of an entire galaxy of stars. Given such astounding luminosity, Swedish astronomer Knut Lundmark provided these objects with the name that stuck—*supernovae.*

Soon after Baade transferred to the Mount Wilson staff in 1931 to use the world's largest telescopes to pursue his observations of supernovae (among a bevy of other interests), physicist Fritz Zwicky from the nearby California Institute of Technology joined up with him as a collaborator. Born in Bulgaria of Swiss parents in 1898, Zwicky was educated in Zurich and remained a Swiss national all of his life. He had arrived at Caltech in 1925 as a research fellow in the physics department, where he studied the physical properties of liquids and crystals. But that was just for starters. Restless in his interests, he eventually rose to a professorship and published over his lifetime nearly six hundred scientific papers, with a sweeping range of topics: cosmic

Fritz Zwicky (*Photograph by Fred Stein, courtesy of the American Institute of Physics Emilio Segrè Visual Archives*)

rays, the extragalactic distance scale, age of galaxy clusters, aerial propulsion, meteors, ionization of gases, quantum theory, elasticity in solids, crystal lattices, electrolytes, gravitational lensing, propellants, and quasars.

Despite Caltech's relaxed campus atmosphere, a hallmark of the California lifestyle, Zwicky retained the authoritative air of a nineteenth-century European professor. He was an aggressive, original, and stubbornly opinionated man, the supreme scientific individualist. He regularly annoyed his physics and astronomy colleagues by studying anything he pleased (he called astronomy his "hobby") and championing along the way some pretty wild ideas, some that waited decades to be proven true. In 1933 he was the first to propose, for example, the existence of cosmic "dark matter" (what he called in German "dunkle Materie"), today one of astronomy's outstanding mysteries. "Zwicky was one of those people," recalled Caltech astronomer Wallace Sargent, "who was determined to show the other guy was wrong. His favorite phrase was, 'I'll show those bastards,'" which he did to the fullest.

Baade and Zwicky were astronomy's odd couple: where Zwicky was cranky, imperious, and for the most part a solitary researcher, Baade was soft-spoken, even-tempered, and a team player. Yet, sharing a common language and cultural heritage, the two became fast friends and could often be seen about town talking endlessly about novae (until they had a terrible falling out years later).

Some of their best work together took place in 1933. While Chandrasekhar in England was reluctant to speculate on what might happen to a star more massive than the typical white dwarf star, Zwicky was quick to offer a suggestion. Just the year before in Great Britain, James Chadwick had bombarded atomic nuclei with high-energy radiation and succeeded in wrenching out some particles, which had all the properties of a particle that theorists were suggesting might exist. Each had just about the same mass as a proton but with no electrical charge. This particle was neutral—hence its name, the neutron.

On hearing this news from the world of particle physics, Zwicky, in his usual madcap way, figured he could use this newfound particle to explain how a supernova ignited. Somehow (Zwicky didn't yet know the exact route), a stellar core would get squeezed and squeezed over time, until it reached a tremendous density, just like that of an atomic nucleus. The negatively charged electrons and positively charged protons in the stellar core, in this event, would be pressed inward to form a naked sphere of neutrons. "Such a star," he and Baade wrote in the *Proceedings of the National Academy of Sciences,* "may possess a very small radius and an extremely high density." In fact, a width not much longer than a city. It seemed natural for Zwicky to call it a "neutron star."

Since Zwicky's day, astronomers have now worked out that route to a supernova. It all depends on how massive the original star is. An average star over its lifetime carries out an amazing balancing act. Gravity is continually pulling the matter inward, trying to squeeze it down tighter and tighter. But at the same time the tremendous pressure of the star's hot gases pushes outward. What results is a stable star emitting light and energy into the universe. Our Sun has walked this tightrope for some five billion years and will do so again for another five billion. But there's an end to this road. The hydrogen atoms being fused into helium are eventually exhausted, gravity takes over, and the core shrinks. With this gravitational energy liberated, the outer envelope of the star expands outward and cools, thus creating a giant star, no longer yellow but now a cooler red. At this point, helium takes over as the fuel.

For our Sun the helium, too, will ultimately be used up, yet the nuclei will fuse further—generating carbon and oxygen. But that is the Sun's endpoint. The Sun is not massive enough to fuse those atoms into heavier elements. It will run out of fuel. When that happens, its red giant envelope will eventually whisk away, and what is left—the hot core—will remain behind as a white dwarf star, about as big as the Earth. With the nuclear engine turned off, this stellar nugget, around three-fifths the mass of the present-day Sun, will begin slowly to cool. A spoonful of Earth has an average density of five grams per cubic centimeter. In a white dwarf, it can range from ten thousand to a hundred million grams per cubic centimeter. Gravity is the ultimate cosmic vise, capable of packing the mass of a high-rise building into the space of a sugar cube. Ultimately, "electron pressure" keeps the white dwarf from compacting further. The electrons don't budge; they're still powerful enough to hold gravity at bay.

But what happens if the star is more massive than the Sun? First of all, the star can continue to burn beyond carbon and oxygen. The atoms fuse into neon and magnesium; these in turn serve as the raw materials for constructing even heavier elements, such as silicon, sulfur, argon, and calcium. If the star is massive enough, this can go on and on until ultimately iron is formed. But that's the end of the line. The star's terminus. Its final depot. Fusing iron takes more energy than it will release. So at this critical moment the star faces its waterloo. Unable to generate any more energy, gravity takes charge. In fact, it enters the scene like gangbusters. As soon as a stellar core turns to iron, it collapses catastrophically. In less than a second, a core that was once the size of the Moon is squeezed down to the size of a city. That's right—in less than a second.

How could this be? It's because the electrons can no longer hold up against gravity. In the course of the titanic stellar collapse, each electron ends up merging with a proton to form a neutron, a neutral particle, releasing a flood of tiny neutrinos in the process. What forms is a neutron star about a dozen miles (twenty kilometers) wide. This sphere is so dense that it's essentially one humongous atomic nucleus, more than 100,000,000,000,000 times denser than the Earth. (It was once estimated that your bathroom sink could hold all the water of the Great Lakes if the water were compressed to the density of a neutron star.) If a mountain did form on the neutron star, it couldn't get any higher than a few centimeters, given the strength of its gravitational field. In this situation it is now the *strong* nuclear force that begins to play a role in holding the star in place, resisting the pull of further gravitational squeezing. Nuclear forces are a major defense against the relentless tug of gravity to make the star even smaller.

The evidence for all this is in the grand announcement of this event: the shock wave sent out from the collapse, along with the torrent of neutrinos, speeds through the remaining stellar envelope. When they emerge at the surface, we see the result as a spectacular explosion—a supernova, the birth announcement for the neutron star, just as Zwicky had predicted in the 1930s. And in the process, elements beyond iron are forged within the chaotic turbulence of the explosion's cloud.

Zwicky, of course, did not know all these details at the time. He just figured that the supernova was fueled by the tremendous energy loss as the stellar core got smaller and smaller, somehow releasing that power in a stupendous burst. But his embryonic vision of the process was still an astounding and prophetic prediction; the neutron star wouldn't be confirmed for another three decades. In Zwicky's day, neutron stars remained theoretical fabrications, which astronomers figured would never be seen even if they did exist, due to their extremely small size. (That all changed when the first bona fide neutron star, beeping away as a radio pulsar, was at last discovered by the British astronomer Jocelyn Bell in 1967.)

Baade and Zwicky first presented their prescient ideas at a meeting of the American Physical Society at Stanford University in December 1933. Astronomers liked the idea of the supernova—an extra powerful stellar explosion—but considered the concept of the neutron star far-fetched and wildly speculative. They figured that supernovae allowed extra-massive stars to eject enough mass to settle down as white dwarfs. Astronomers had barely gotten comfortable with the idea that matter could be crushed to huge densities within a white dwarf star. As a consequence, hardly anyone took the neutron star seriously—except for a few brave souls. Chandra remarked on the possibility at a Paris conference in 1939, agreeing that the formation of a neutron core "may be the origin of the Supernova." But he didn't immediately follow up. Those who did included Lev Landau in the

Soviet Union and, at the University of California, Berkeley, J. Robert Oppenheimer, who went on to become the father of the atomic bomb. That these two physicists didn't dismiss the neutron-star idea right off, but instead pursued it, was a crucial turning point for the black-hole story. It led researchers, at least a few at first, to suspect that the cosmos might well be generating those danged singularities.

6

Only Its Gravitational Field Persists

It was a bleak and despairing time for the Soviet Union in the late 1930s. The Great Purge initiated by Joseph Stalin was at its zenith, and Lev Landau figured he was in the crosshairs, despite being a fervent Marxist. When rules for Soviet scientists were looser in the 1920s, Landau had spent time in Western Europe, visiting the top university centers in physics. He was the topsy-turvy-haired wunderkind whose research articles on a wide range of physics problems were noted for their creative insight and mathematical dexterity. Landau's genius was recognized by everyone who met him. But soon after he returned in 1931, it became a crime for Soviet scientists to maintain any contacts with the West, for fear of capitalist contamination. Just having visited the West—even in an earlier, more liberal era—made Landau suspect.

By 1937 the purge was reaching beyond the Communist Party into the intelligentsia. As a result Landau, twenty-nine years old at the time, decided to put his current work on atomic physics, magnetism, and superconductivity on hold and to take another look at the problem of stellar energy, hoping for a breakthrough in one of physics' greatest challenges. An astute practitioner in the art of academic politics, Landau reckoned the scientific glory that was bound to come if he figured out how a star was powered might protect him from getting arrested. Many of his colleagues had already been caught up in the sweep. The worldwide attention he'd receive for a scientific triumph, he believed, would force officials to spare him.

Striving for a completely new approach, one that didn't involve astrophysicists' standard stellar models, Landau arrived at the conclusion that stars have a "neutronic core." While Zwicky considered that the formation of a neutron star was the trigger for a supernova, Landau concluded that atomic nuclei and electrons combine to form neutrons in the dense heart of every *normal* star. According to Landau, this made the core more compact, which liberated enough energy to power a star over eons. Landau's friend the physicist George Gamow, in a book he published that year on atomic physics, calculated that the density of such a stellar core would be around a hundred trillion grams per cubic centimeter, "analogous to the conditions inside an atomic nucleus." The gravitational energy liberated as the core squeezed down to such an immense density, Gamow went on, would "be quite enough to secure the life of the star for a very long period of time."

Lev Landau (*American Institute of Physics Emilio Segrè Visual Archives, Margrethe Bohr Collection*)

Landau mailed his manuscript to Niels Bohr in Copenhagen. As Bohr was an honorary member of the Soviet Academy of Sciences, this was still an allowable route for getting work noticed in the West. Bohr passed it along to the scientific journal *Nature*, which published the paper on 19 February 1938. In it, Landau claimed that "we can regard a star as a body which has a neutronic core the steady growth of which liberates the energy which maintains the star at its high temperature."

Once in print, Landau shrewdly devised a public relations campaign to spread the word. Through contacts in high places, he got one of the USSR's most influential newspapers to praise his scientific paper, which described it as a "bold idea [that] gives new life to one of the most important processes in astrophysics."

It was a good try—but not good enough. Landau's political strategy—Bohr's backing, the glowing press coverage, publication in a prestigious journal—failed miserably. (That he had prepared an anti-Stalinist leaflet to hand out during 1938's May Day parade likely contributed to the outcome.) Landau was eventually arrested and imprisoned for a year on the ludicrous charges that he, a Jew, had been spying for Nazi Germany. Not until the intervention of the noted Soviet physicist Pyotr Kapitsa, who stridently told Soviet authorities that only Landau could explain a newly discovered phenomenon called superfluidity, was he released. And Kapitsa was right. Landau did come to solve how some supercooled liquids can flow without friction, for which he received the Nobel Prize in 1962.

Although Landau proved brilliant on superfluids, he failed on stellar energy. His physics in that arena was seriously flawed. The very next year in 1939, the German-American physicist Hans Bethe cracked the mystery of how the Sun shines. He was the first to devise a plausible pathway for stars to generate their immense energies through the fusion of atoms rather than the release of gravitational energy.

Still, Landau's *Nature* article was highly influential in advancing the story of the black hole. The paper arrived at the desk of Caltech theorist Richard Tolman, a world expert on general relativity, who enthusiastically embraced the idea of a neutron star. Tolman saw it as a problem crying out to be solved, and he urged his colleague J. Robert Oppenheimer to apply Einstein's space-time equations to collapsing stars. Oppenheimer was already aware of Landau's model of high concentrations of neutron matter and intrigued by it. Encouraged by Bethe, who was visiting Berkeley, Oppenheimer worked with Robert Serber in the summer of 1938 to check Landau's paper out. They quickly figured that normal stars, such as the Sun, could not possibly harbor neutronic cores; otherwise, they would look very different. The Sun, for example, would be far smaller, owing to the huge gravitational pull of an ultradense center.

Even though Landau's central idea failed when it came to explaining stellar power, Oppenheimer was inspired by its description of dense stellar cores. If they couldn't resolve how stars shine (Bethe took care of that), Oppenheimer began to wonder whether Landau's neutron cores played a role at the *end* of a star's life. Could Zwicky be right after all?

Like any good physicist, Oppenheimer reduced the problem to its essentials. He ignored any talk of stars blowing up. To Oppenheimer, people like Zwicky, a physicist he did not greatly admire, did that sort of grandstanding. Oppenheimer focused solely on the neutron star itself. What was its physics? Chandrasekhar had discovered that a white dwarf star needs to stay under a certain mass before it transforms into something else: Did a neutron star have a similar limit? Though Oppenheimer dealt with these stellar questions only briefly in his professional life, they led to some of his greatest achievements in theoretical physics, a choice of fields one might not have expected based on his family background.

Oppenheimer had grown up on the Upper West Side of New York City amid privilege and comfort. His father had secured the family's wealth from the textile trade. As a young boy, Oppenheimer was driven to private school by a uniformed chauffeur in a limousine. A solitary child (a younger brother didn't arrive until he was eight), young Robert was particularly fascinated by rocks and minerals, and he filled the family's Manhattan apartment with specimens.

At Harvard as an undergraduate, he was first drawn to chemistry. Finding himself inept at experimental lab work, a failing he freely acknowledged, he was soon attracted to more theoretical pursuits, especially in physics. Quantum mechanics was revolutionizing physics in the mid-1920s, and Oppenheimer, eager to jump on the bandwagon, arranged to do his graduate work in Europe, first at Cambridge University in England and then at the University of Göttingen in Germany, becoming acquainted with the field's all-stars, among them Paul Dirac, Niels Bohr, and Max Born.

Although Oppenheimer missed the first wave of quantum mechanical revelations, the sort that led to bevies of Nobel Prizes, he did become immersed in the second wave: the attempt to join special relativity theory to quantum mechanics. Dirac's prediction that "antimatter" existed came out of these explorations.

After obtaining his doctorate, Oppenheimer returned to the United States just as business and government were pushing to enhance graduate science programs in the nation's universities. With his golden European credentials, Oppenheimer was a hot prospect, and in 1929 he secured a joint position with the California Institute of Technology and the University of California at Berkeley. What he ultimately established on the West Coast through the 1930s and into the early 1940s was one of the best "schools" for theoretical physics in the world, with the most gifted students in the field flocking to both campuses to work with him. He and his students dealt with either the new particles and forces revealed by Ernest Lawrence's cyclotron at Berkeley or the astronomical and astrophysical discoveries unveiled by Caltech professors. "What really made this school a success was Oppenheimer himself—not so much for his brilliant physics or his classroom skill or his administrative maneuvering, none of which was of the first rank, but for his unique, European-acquired ability to select the most promising problems for his group and to inspire and guide his charges to the leading edge of research in these areas," explains science historian David Cassidy.

An acquaintance of Oppenheimer's at Berkeley said that students considered the physicist "something like a god," especially given his

J. Robert Oppenheimer (*Los Alamos Scientific Laboratory, courtesy of American Institute of Physics Emilio Segrè Visual Archives*)

tall stature, arresting blue eyes, wealth, and extensive interests outside of science, including painting, Sanskrit, and reading Plato in the original Greek. He was a charismatic figure, generous in sharing credit with his students. Yet, he was also a man haunted by personal demons and insecurities.

Oppenheimer was not a mind-blowing, theoretical innovator—not in the way others, such as Werner Heisenberg, broke radically new ground in modern physics. There is no Oppenheimer Uncertainty Principle. He did work with Max Born to forge the Born-Oppenheimer approximation, still used to calculate the quantum behavior of molecules, but Oppenheimer mainly specialized on calculations that could explain ongoing atomic physics experiments—yeoman work that was necessary and vital but did not often gain the spotlight. Many of his theoretical papers are now out-of-date or largely forgotten. It was the fleeting detour that he took at the end of the 1930s into the world of astrophysics that generated the scientific papers that are most referred to today. What he and his students did was tug the Schwarzschild singularity into the real world.

It's not surprising that his work in atomic and nuclear physics theory would draw him into the cosmic arena. As noted earlier, in the 1930s physicists were struggling to explain how the Sun and stars could be powered over billions of years, and it was already apparent that the source had to be a nuclear process that released energy as atoms combined. A number of physicists had made note of this possibility a decade earlier. The question was the precise pathway. Landau had proposed an entirely different scheme altogether, primarily because theorists were encountering so many obstacles in showing how atoms could fuse within the Sun. So much so that Landau was convinced that the physics done by leading astrophysicists, such as Eddington, was irrational.

For years, Oppenheimer kept his eye on the field and eventually helped organize a symposium on the astrophysical significance of nuclear transformations at the 1938 joint meeting of the American Physical Society and the American Association for the Advancement of Science. But at that very moment Hans Bethe at Cornell University was completing his historic paper that revealed the first bona fide route for a star to fuse its hydrogen into helium and generate nuclear energy, an achievement that led to Bethe's receiving the Nobel Prize in 1967.

Seeing that he was scooped, Oppenheimer turned his attention to the other end of a star's life—its death. By then Fritz Zwicky had suggested that a star blowing up would leave behind a dense ball of neutrons, as the protons and electrons in the collapsing core were crushed together. Landau at the same time was talking of stars having "neutronic cores." Chandra had required only special relativity to arrive at his limit for the white dwarf star. But the neutron star was an arena where general relativity was necessary to get answers. With the neutron star's high density and intense gravitational field, Newtonian laws of gravity were no longer adequate. To carry out this endeavor, Oppenheimer joined forces with his graduate student George Volkoff (with Tolman consulting from the sidelines). The pair worked out a full, general-relativistic treatment of how such a neutron core might form. In this precomputer era, Volkoff labored over a calculating machine to carry out the intricate number crunching. In the end, they proved that neutron stars, in all probability, *were* inhabiting the universe. Zwicky was right. But you wouldn't know that from reading Oppenheimer's writings on the topic. The very title of his 1939 *Physical Review* paper with Volkoff refers, not to neutron stars, but to "neutron cores," favoring Landau's description. And Oppenheimer made sure to not cite anything by Zwicky. It is Landau's work that is generously credited throughout the eight-page paper.

Zwicky, hearing of their work, was furious that they were not conferring with him. The prickly physicist figured he was the world's expert on this topic. By then he was carrying out systematic searches for supernovae, his suggested birthplace for neutron stars, using a special large-field telescope on California's Palomar Mountain. He fired back at Oppenheimer with a pedestrian article that same year titled "On the Theory and Observation of Highly Collapsed Stars" in the same journal, with not one reference to Oppenheimer and Volkoff's groundbreaking paper. That was Zwicky's tit for Oppenheimer's tat. Zwicky's paper is largely forgotten today. He had the chutzpah and courage to speculate—properly getting the credit for imagining a neutron star emerging from the spectacular explosion of a star—but it was Oppenheimer and his brilliant student Volkoff, notes Caltech theorist Kip Thorne in his book *Black Holes and Time Warps,* who had the theoretical savvy to be the first to master the physics of this strange new stellar object. Thorne calls their paper "a tour de force, elegant, rich in insights, correct in all details."

This pioneering paper was also titillating in its final conclusion. Volkoff and Oppenheimer found that there was an endpoint to the neutron star: past a certain mass, the neutron core would continue to contract—and contract indefinitely. Just as Chandrasekhar found a limit for a white dwarf's size, Volkoff and Oppenheimer revealed a similar constraint for the neutron star. What becomes of the star if its mass steps past the boundary? "The question of what happens . . . remains unanswered," they replied. But no one was panicking as yet. They knew they were just getting started on this problem. It was still possible that the physics of such condensed matter was not fully understood; perhaps new repulsive forces come into play to prevent the ultimate collapse.

To find out, Oppenheimer recruited another graduate student, Hartland Snyder, who had a reputation as a crackerjack mathematician and could handle general relativity with ease. It was a unique pairing. Snyder came from the working class. And "Oppie was extremely cultured; knew literature, art, music, Sanskrit. But Hartland—he was like the rest of us bums. He loved the . . . parties, where . . . we sang college songs and drinking songs. Of all of Oppie's students, Hartland was the most independent," Caltech physicist William Fowler once recalled. Oppenheimer asked Snyder to take the story further, to find out what happens to that collapsing neutron star, the one that goes past the limit. And the results of this endeavor, Oppenheimer later told a colleague, were "very odd."

Oppenheimer and Snyder began with a star that has depleted its fuel. And to make the computation easier in that era of clunky, desktop calculating machines, they ignored certain pressures and the star's rotation. Otherwise, the problem would have been impossible to solve.

With the heat from its nuclear fires gone, the star's core cannot support itself against the pull of its own gravity, and the stellar corpse begins to shrink. Oppenheimer and Snyder determined that if this core is weightier than a certain mass (now believed to be around two to three solar masses, the kind of cores found in massive stars weighing twenty-five Suns or more), the stellar remnant would neither turn into a white-dwarf star (our own Sun's fate) nor settle down as a ball of neutrons. That's because once the material is squeezed to densities beyond four hundred billion pounds per cubic centimeter, the neutrons can no longer serve as an adequate brake against collapse. Degeneracy pressures, this time from neutrons, no longer do the job. Oppenheimer and Snyder calculated that the star would continue to contract indefinitely. There is no rest for the weary when gravity takes over. The matter within such a collapsing star is in a state of permanent free fall.

The last light waves to flee before the "door" is irrevocably shut get so extended by the enormous pull of gravity (from visible to infrared to radio and beyond) that the rays become invisible and the star vanishes from sight. Space-time is so warped around the collapsed star that it literally closes itself off from the rest of the universe. "Only its gravitational field persists," the Berkeley physicists reported.

They figured that the star collapsed to a point, a singularity squeezed to infinite density and zero volume (which seems impossible). Their equations indicated this, but they hesitated in saying it directly. That's because singularities are a horror to physicists. It is a signal that something is wrong with the theory under these extreme conditions, that they had entered a realm where the mathematics being used ceases to be a valid description of the physics. It's as bad as trying to divide a number by zero. How many zeroes are there in the numbers 8, 29, or 103? There is, of course, no definitive answer. There is a countless number of zeroes; in other words, 29 + 0 + 0 + 0 + 0 + (an endless series of zeroes) still equals 29. It's a mathematical operation that leads nowhere. Zero divided into 29 equals infinity, which is not a satisfying answer. A singularity in a physics equation, where a parameter flies off to infinity, signals a similar breakdown. Given this predicament, Oppenheimer and Snyder were willing to go only so far. What they did report was bizarre enough. Werner Israel has called this "the most daring and uncannily prophetic paper ever published in the field. . . . There is nothing in this paper which needs revision today."

In the title of their paper, Oppenheimer and Snyder called this phenomenon "continued gravitational contraction," and it established the first modern description of a black hole. But few became aware of it, partly due to unfortunate circumstances. Oppenheimer and Snyder published their paper in the *Physical Review* on 1 September 1939, the day Hitler ordered his troops into Poland, triggering the start of World War II. No wonder it received little notice. More than that, the same journal issue held a seminal paper by Niels Bohr and John Wheeler on nuclear fission, then a far more urgent topic on physicists' minds. Collapsing stars seemed of little import by comparison. It was the last paper that Oppenheimer wrote on the subject. With no physics yet developed to follow the collapsing matter into its abyss, what more could he do?

Professionally, it was a transitory detour in Oppenheimer's scientific life, involving only three papers. Afterward, he went back to his work on nuclear particles and cosmic-ray physics and, by 1942, into the nation's Manhattan Project aimed at manufacturing the world's first atomic bomb. His students, after graduation, went off to university teaching positions, never returning to the topic. Most astronomers, if they thought of this problem at all, assumed that massive stars got rid of most of their mass over time—enough to safely keep them as white dwarf stars in old age. Only Fritz Zwicky kept banging the drum about neutron stars, publishing a few papers on the topic. No one took note.

Perhaps stellar winds, said astronomers, carried off a lot of an old star's mass into space; maybe stellar explosions kept any stellar remnant at one solar mass or less. This was not an unreasonable assumption, for astronomers were just coming to recognize Wolf-Rayet stars, which do just that. These evolved and weighty stars eject tremendous amounts of mass each year, a billion times more than our own Sun, via strong stellar winds.

And even if a celestial object did gravitationally collapse somewhere in the heavens, it was still in essence invisible. No telescope at the time was capable of confirming the existence of a neutron star or black hole. They didn't know it at the time, but astronomers had to wait for the development of new tools and new techniques for scanning the heavens—to capture electromagnetic waves *beyond* the visible spectrum.

And what about the general relativists? Weren't they excited by this new and astounding finding coming out of general relativity? In truth, they weren't even paying attention. Relativity experts at the time were more interested in the esoterics of curved space-time, not its practical use in astrophysics. The general theory of relativity was then virtually a playtoy for mathematical physicists, fun to delve into but unconnected in their minds to the celestial sky (except, maybe, for the bending of some starlight near the Sun). General relativity at the time was primarily taught as mathematics, not physics. They were interested not in experimental evidence or applications but rather in rigorous proofs.

Even though no one followed up on the gravitational collapse in the West, Landau in the Soviet Union was mightily impressed. He added the Oppenheimer-Snyder paper to his "golden list," a running tally of classic papers he considered worth checking out.

Many years later physicist Freeman Dyson tried to talk to Oppenheimer about his work with black holes, but the father of the atomic bomb would have none of it. Because Oppenheimer believed that he was merely applying Einstein's laws to collapsing stars and not unearthing a new law of physics, Dyson suspects that Oppenheimer thought of his accomplishment as worthy only of "graduate students or third-rate hacks." He didn't recognize his deed as a theoretical triumph. But Dyson strongly disagreed. He described Oppenheimer's paper on continued gravitational contraction as his "most important contribution to science . . . a masterpiece of derived science, taking some of Einstein's basic equations and showing that they give rise to startling and unexpected consequences in the real world of astronomy."

Yet as the 1930s were coming to a close, most astronomers were not yet ready to believe that such bizarre objects could be generated in the real world. Even Einstein wrote a paper in 1939, published a month after Oppenheimer and Snyder's, attempting to prove they were impossible to form. His calculations would have arrived at the same answer as the California theorists, except that Einstein stacked the deck and arranged his model in such an unrealistic way that his star could never collapse. "One could not be sure," wrote Einstein, "that . . . assumptions have been made which contain physical impossibilities." Okay, he was saying, you can create a singularity on paper mathematically, but matter likely acts in such a way to thwart the collapse. Einstein tried to prove this by imagining the mass as a "great number of small gravitating particles . . . resembling a spherical star cluster." It was a sleight of hand in which the centrifugal force of the particles' circular motion essentially keeps them from all collapsing to a singular point. And it was just that—an illusion. Infamous for not keeping up with the scientific literature, Einstein had not read Oppenheimer and Snyder's paper before tackling the problem on his own.

Some historians have labeled Einstein's 1939 disproof of the singularity as a "strong candidate for the dubious distinction of being his worst scientific paper." That's because Oppenheimer and company were spot on. Once a collapsing star gets small enough, nothing in the universe can stop gravity from creating a black hole—no amount of rotational motion or gas pressures. Gravity is the ultimate trump card, overwhelming any internal stellar forces. Why couldn't Einstein recognize this simple physical fact? Because, says Caltech general relativist Kip Thorne, Einstein "was so firmly convinced they cannot exist (they 'smelled wrong'; terribly wrong) that he had an impenetrable mental block against the truth—as did nearly all his colleagues." It was the "mindset of nearly *everybody* in the 1920s and 1930s," he says. In their heart of hearts, most physicists wanted to uncover a law of physics that forbids black holes from forming. Trained in Victorian times, these scientists had to step over formidable psychological hurdles before accepting something so unexpected in nature.

"There is a curious parallel between the histories of black holes and continental drift," Werner Israel has noted. "Evidence for both was already non-ignorable by 1916, but both ideas were stopped in their tracks for half a century by a resistance bordering on the irrational." Israel blames the threat each concept posed to our cherished faith in the permanence and stability of matter. Whole continents wandering about the Earth like chess pieces? Stars disappearing from space and time? Surely that must be poppycock!

Given today's thriving interest in an Einsteinian universe, it's now difficult to appreciate this line of thinking. But this dismissive attitude came about during a period when the theory of general relativity was pushed into the shadows, admired from afar for its mathematical beauty but largely ignored. Theorists revered Einstein's equations (almost as sublime mathematical sculpture), but they were not actively engaged in working with them in any way. That was especially true in Germany once the Nazis rose to power. As part of their campaign against "Jewish physics," they officially forbade relativity to be taught throughout the Third Reich. But, politics aside, universities around the world rarely offered general relativity as a course, and if they did, it was taught as mathematics, not physics. Most theoreticians at the time were more focused on quantum theory, with its new and revolutionary perspective on matter and energy. And outside the tight-knit world of theoretical physics, Einstein's theory of gravitation was actually unpopular. "It was despised, occasionally even abhorred, by the specialists in other fields of physics," says physicist and historian Jean

Eisenstaedt. This is because it dealt "with some tricky concepts that the ordinary physicist finds difficult to understand." It requires us to think of space and time in a way that fights against our everyday experiences of how the world works—"probably also our brain processes," notes Eisenstaedt.

Moreover, the myth had arisen soon after general relativity's introduction that less than a handful of people truly comprehended it. Arthur Eddington himself liked to tell the tale of being approached at a Royal Society meeting and told, "Professor Eddington, you must be one of three persons in the world who understands general relativity." When Eddington hesitated, the questioner continued, "Don't be modest." To which Eddington replied, "On the contrary, I am trying to think who the third person is."

Some believe this exaggerated impression of general relativity's difficulty kept people away and contributed to the field's stagnation. General relativity nearly withered on the vine. After a flurry of scientific publications in the early 1920s, after the theory's well-publicized verification in 1919, interest plummeted. For some three decades afterward, less than 1 percent of physics journal articles involved relativity. It was once said at a conference that you could count the number of general relativists in the world "on the fingers of one hand." Oppenheimer's startling new contribution hardly made a difference in the ongoing diminution of the field. Oppenheimer himself, after he became head of the Institute for Advanced Study in Princeton, New Jersey, in 1947 advised the up-and-coming physicists there to avoid pursuing general relativity, believing it a dead end. Einstein, then in his waning years, worked in an office just down the hall from Oppenheimer.

Ultimately, physicists want a theory to connect with the world. You could argue that quantum mechanics was just as weird—what with particles acting as waves, waves as particles. Why was it so readily accepted and general relativity snubbed? It's primarily because quantum theorists worked hand in hand with experimentalists. There was a deep pool of experimental data to support quantum mechanics' predictions on the nature and behavior of matter (weird as it was) on very small scales. In the late 1920s Paul Dirac, for example, posited the existence of antimatter, and by 1932 an experimentalist found evidence for this new type of particle, bizarre as it sounded, in a cosmic-ray bubble chamber. General relativity, on the other hand, had merely a wobble in Mercury's orbit and some bending of starlight skirting past the Sun to back up its tenets. The noted physicist Richard Feynman wasn't a fan of general relativity for that very reason. "Because there are no experiments this field is not an active one, so few of the best men are doing work in it . . . ," he wrote his wife from a world conference on gravity. "The good men are occupied elsewhere."

After the brief but seminal flurry of work coming out of Oppenheimer's group at Berkeley, the subject of total gravitational collapse didn't just get placed onto the backburner; it got shoved into a closet. World War II only accelerated the process. It sidetracked many physicists into the more vital concerns of the moment: radar, nuclear physics, military technology. "We who worked in this field," said Leopold Infeld, who collaborated with Einstein, "were looked upon rather askance by other physicists. Einstein himself often remarked to me: 'In Princeton they regard me as an old fool.' This situation remained almost unchanged up to Einstein's death."

7

I Could Not Have Picked a More Exciting Time in Which to Become a Physicist

It was not until the mid-1950s that interest in general relativity and its applications was at last revived after its decades-long lull. It was in the nick of time. The field had gotten so moribund that the Dutch American physicist Samuel Goudsmit, codiscoverer of electron spin and then editor-in-chief of the *Physical Review*, was about to lay down an edict that papers on general relativity would no longer be accepted by the journal. But by that point a renaissance for general relativity was beginning to bloom in the Soviet Union, Europe, and the United States. This happened for a variety of reasons. For one, the start of the space race and the Cold War led to more funding, especially in the United States; given their experience in the war, the US military branches had learned of the great advantages in sponsoring basic research in all types of fields. The worldwide celebration of special relativity's fiftieth anniversary in 1955 also brought many physicists together, allowing them to recognize that gravitational research had been unfairly neglected and was due for more attention. Barely alive within the journals of physics through the Depression and the war, the topic slowly gained more and more significance. "Within a few years, understanding of gravitational collapse progressed from its inchoate beginnings to a sophisticated discipline," said theorist Werner Israel.

One of the more unusual reasons for gravitation's reawakening can be traced to an eccentric American financier. Born in Gloucester, Massachusetts, in 1875, Roger Babson graduated from the Massachusetts Institute of Technology with an engineering degree but was lured into the stock market during its glory days in the 1920s, where he applied his statistical training at the brokerage firm he established. Such a mathematical approach on Wall Street was fairly novel at the time, but the MIT graduate had a physics mind-set. "Babson was fascinated with Newton's three laws of motion, and sought to apply them directly to his own studies of business trends. Most important was Newton's third law," science historian David Kaiser has written. That law states that for every action there is an equal and opposite reaction, which to Babson meant that the high-flying stocks then in play would surely plummet someday. Shortly before the market's dramatic fall in 1929, he sent out a forecast to his clients of the imminent collapse and placed his money in safe havens, allowing him to sail "through the Great Depression as one of America's wealthiest citizens," noted Kaiser.

Convinced that Newton saved him from financial ruin, Babson and his wife began to gather a vast collection of Newton's original publications, as well as books that the great Sir Isaac owned. The couple went so far as buying the entire front parlor of Newton's London home (including its walls) and setting it up in a special room at the institution he founded, Babson College in the suburbs of Boston, where it remains today.

These enthusiasms for all things Newton eventually led Babson to set up and completely fund the Gravity Research Foundation in 1948. He had come to believe that gravitation was a specialty that deserved more attention among physicists, and the foundation began generously subsidizing conferences on gravity and establishing lucrative annual awards for the best essays written on gravity. Although his largesse provided a real boost to gravity researchers, Babson's true goal was to find a way to conquer gravity. He hoped that his funds would lead to the development of "antigravity," a means to counteract the attractive force. It was an obsession for Babson ever since his oldest sister drowned when he was young; he blamed gravity for pulling her to the water's bottom. Special insulators and shields stop magnetism, why not search for comparable insulators that could vanquish gravity, thought Babson. When the Tufts University physics department received a hefty grant from Babson's foundation in 1961, it became the owner of a large stone monument in honor of this pursuit. Carved into the rock are these words: "It is to remind students of the blessings forthcoming when a semi-insulator is discovered in order to harness gravity as a free power and reduce airplane accidents." (Similar stones, about a dozen in all, were donated by Babson to other colleges situated in New England, the South, and the Midwest.) "[Tufts] legend has it that from time to time, groups of fraternity brothers band together to move the 2,000-pound monument to a different location during the night, working like anti-gravity's little elves," said Kaiser.

The foundation's essay prize was at first looked upon as a joke, due to its initial focus on antigravity, leading some to label those working on gravitation as "mad men and quacks." But the tide turned when a young relativist named Bryce DeWitt, in need of a down payment for a house, submitted a paper that brazenly argued how searches for gravity reflectors or insulators were "a waste of time." DeWitt laid out more reasoned arguments to study gravitation and won the prize. As a result, the contest eventually attracted entries from talented physicists in gravitational research, and the submissions duly shifted to broader concerns in relativistic theory. (The contest continues, with the era's most noted theorists, among them Stephen Hawking and Roger Penrose, winning its essay awards.)

The stone placed by the Gravity Research Foundation on the campus of Tufts University in 1961. (*Daderot, courtesy of Wikimedia Commons*)

In the 1950s the foundation's president convinced another rich industrialist, Agnew Bahnson, to support a new institute for gravitational studies at the University of North Carolina, to be headed by DeWitt, a pioneer in the quest to join quantum mechanics with general relativity. To separate themselves from the "mad men" and make sure the physics community deemed them legitimate, institute physicists openly declared in their literature that they had "no connection with so-called 'anti-gravity research' of whatever kind and for whatever purposes." Within months of its founding, the new institute in early 1957 held a conference on the role of gravitation in physics, a meeting now deemed a "landmark" in the rebirth of gravitational studies.

"By organizing conferences, sponsoring the annual essay contests, and making money and enthusiasm widely available for people interested in gravity, the eccentric Gravity Research Foundation may claim at least some small amount of the credit for helping stimulate the postwar resurgence of interest in gravitation and general relativity," says Kaiser.

In the United States the epicenter of this revival was situated at Princeton University, where physicist John Archibald Wheeler had decided to ponder the fate of collapsed stars, taking up where Oppenheimer left off. It was Wheeler, furiously working behind the scenes, who got Goudsmit to reverse his proposed ban on general-relativity articles in *Physical Review*. Wheeler spent most of his academic life at Princeton, where he set a record in the number of graduate and undergraduate students he supervised—nearly a hundred, including Richard Feynman. As a young man, Wheeler did groundbreaking work in nuclear physics but later made his greatest contribution in science by almost single-handedly taking general relativity, a field that had been stagnating for decades, and applying it to the universe at large.

Under Wheeler's expert guidance, the subject was reborn, as he inspired his small army of students and post-docs to seek ingenious solutions that could be meaningful in understanding the cosmos. As Wheeler put it, he wanted to usurp the " 'one-legged men'—men who knew nothing *but* relativity!" He desired to take the theory out of its ivory tower and couple it to real-world observations—to allow his students to stand on *two* legs. When someone mentioned the word *relativist* to him, Wheeler replied, "There is no such thing; they are physicists."

Wheeler's forays into general relativity over the succeeding years led to many research papers that, combined with the notes from his teaching, were turned into a series of noted books on relativity, many in collaboration with his former students. He became America's dean of general relativity. As Freeman Dyson noted on Wheeler's death in 2008, "Before anyone else, he understood that black holes are not merely an odd theoretical consequence of Einstein's theory of gravitation, but must actually exist and play a vital role in the evolution of the universe."

Born in Florida in 1911, Wheeler grew up around the country as his father, a librarian, served at various posts. From childhood, he displayed a knack for mathematics, teaching himself calculus in high school. He also loved machinery, electronics, and explosives, almost losing a finger one day playing with dynamite caps on the family's vacation farm in Vermont.

Admitted to Johns Hopkins in 1927 at the age of sixteen on scholarship, Wheeler first chose to major in engineering. I was "bent on making my own way in the world," he reminisced years later: "Saying 'physics' would have been like saying 'pottery making.'" But the intellectual lure of the new physics then emerging—quantum mechanics, atomic physics, nuclear physics—was too big for him to ignore. "It is no exaggeration to call that period a watershed," he wrote. "Classical ideas about solidity, certainty, stability, and permanence were being abandoned. They were being replaced by *quantum* ideas of uncertainty and granularity and the duality of waves and particles; by *relativistic* ideas of spacetime as cosmic actor, not merely cosmic stage; and by *astronomical* ideas (backed up by relativity) of a universe that is expanding, not static, and of finite age, not eternal. I could not have picked a more exciting time in which to become a physicist."

Wheeler flew through college, not even pausing for a baccalaureate or master's degree. "It was a non stop flight," he liked to say. He went directly from freshman to PhD in six years, completing a dissertation in 1933 at the age of twenty-one on the absorption and scattering of light by the helium atom. Garnering a postdoctoral fellowship, Wheeler then spent time in Copenhagen, known lightheartedly as the "Vatican of physics." There he had the opportunity to meet nearly all the greats in physics, who traveled like disciples to Niels Bohr's institute in the Danish capital to work on nuclear physics with the master. Wheeler's first big splash in the physics community was publishing a paper with Bohr in 1939 on a liquid-drop model of an atomic nucleus, which was crucial to the understanding of nuclear fission and played an important role in the development of the atomic bomb. They had predicted that both uranium-235 and plutonium-239 would be particularly useful in sustaining a chain reaction. Not surprisingly, given this expertise, Wheeler later worked on both the Manhattan Project and the development of the hydrogen bomb. It wasn't until later that general relativity became the love of his life. And when that happened, he said, "I finally had a calling."

He remembered the exact moment when relativity arrived at his door. It was 6 May 1952, at Princeton University, where he had been a faculty member since 1938. The time was 5:55 p.m. That's when he grabbed a new research notebook, bound in black with red leather edging, and wrote down on page 1 both the time and his immediate thoughts in blue flowing ink. (All his professional life, he preferred a fountain pen over a pencil.) Just half an hour earlier, Wheeler noted in his journal, he had learned from the department chairman that he was going to teach relativity, the first time the physics department had offered such a course. He labeled this notebook "Relativity I" and followed up with many others over the years. "I wanted to teach relativity for the simple reason that I wanted to learn the subject," Wheeler later explained. After the war, the fields of nuclear and particle physics

John Archibald Wheeler (*American Institute of Physics Emilio Segrè Visual Archives, Wheeler Collection*)

were in flux. To Wheeler, they appeared "headed toward a complex thicket of pions and countless other particles, and I began to sense that there might be more gold in the general-relativity mine than had yet been unearthed." For one, he had been pondering whether curved space, on the tiniest of levels, might serve as the building material for the elementary particles he had long been studying.

It was a daring yet worthwhile move. Wheeler, as a newcomer to relativity, was able to look at the theory's decades-long problems with unjaded eyes and fresh enthusiasm. He wasn't burdened by the judgments of the past, although he did have some initial prejudices. He had recently come across the classic 1939 papers by Oppenheimer and his students and was terribly bothered by the singularity. Could that really be the fate of a massive star? "I was looking for a way out," said Wheeler. "Something new should happen at the tiniest dimensions, I felt, that would prevent the total collapse. . . . I was convinced that nature abhorred a singularity." It was repugnant to him. His wary attitude toward Oppenheimer may have also played a role: "He seemed to enjoy putting his own brilliance on display—showing off, to put it bluntly. He did not convey humility or a sense of wonder or of puzzlement. . . . I always felt that I had to have my guard up."

Wheeler was not just being stubborn in trying to get rid of the singularity. He sensed that new physics might emerge by tackling this mystery. No one yet understood how gravity acted on the very small scale, at the level of an atom, and here was a means of examining that arena. At the end of its life, a star's core gets squeezed smaller and smaller. What could be learned from this event? Does the matter just disappear, possibly entering another space, another time? Or does it get transformed into a new minuscule state, not yet conceived by our current laws of physics?

Given his background in nuclear physics, Wheeler thought of the proton. From outside the particle, it appears that the proton's electric field is emanating from a point. But in reality the proton has a finite size. Maybe the entire mass of a star was collapsing to a very, very small size, into a state of matter as yet unknown. Or maybe the compacted star fiercely radiates away its mass and energy as it shrinks inward, "until it becomes a cinder too puny for further collapse," mused Wheeler.

That had been the popular escape mechanism for decades. A star, at the end of its life, somehow undergoes a grand fireworks show, ejecting just enough mass to prevent it from ever plummeting to a singular point in the event of a complete gravitational collapse. But to some astrophysicists this belief was "no more than a superstition," a way to avoid confronting the unthinkable.

By teaching a course on general relativity, Wheeler hoped to get better acquainted with the enemy and find the proper means to avoid the ultimate stellar Armageddon. There was good reason for Wheeler's hope: Oppenheimer and Snyder had set up the simplest case possible. In order to be able to handle their calculations, the star in their computations did not rotate; moreover, there were no pressures or shock waves in play. Simply put, it was an ideal star—not a real one at all. What if some force, not yet considered, stepped in to stop a singularity from forming?

Wheeler thought of all the possible escape routes and began testing them out, one by one, mathematically. In one way he was very much like Oppenheimer; he liked to work in close collaboration with his students. He adopted the research style he had learned under Bohr: "free-wheeling talk sessions with colleagues, with more questions than answers flying back and forth, [trying] always to emphasize the positive in my junior colleagues' work, [giving] them all credit that is due," as he put it. And he was highly generous with that credit. Even if he had made the major contribution to a paper in collaboration with a student, the authors were always placed in alphabetical order. That was Wheeler's edict. With his name beginning with "W," that meant his student was highly likely to be listed as "first author," the most honored position in the scientific literature.

Wheeler wanted his students to be bold, like him. Though conservative in his politics and always gentlemanlike in his demeanor, Wheeler was never afraid to dive into the deep end when it came to physics—to try out a range of ideas, no matter how speculative. So much so that he once considered writing a book titled *Not Crazy Enough*. One of his former students, Robert Fuller, notes that this open-mindedness set Wheeler apart as a physicist. He had the ability to dislike a singularity yet be fascinated at the same time by the fact that it appears in the equations. "He was so playful with ideas, so willing to consider the opposite of anything," says Fuller. "He got that from his mentor Niels Bohr, who was always going around saying, 'The opposite of any deep truth is also a deep truth.' Wheeler quoted that every chance he got." He gave any contrary hypothesis its due consideration. Just as Max Planck in 1900 invented the "quantum" to handle a brewing catastrophe in thermodynamics, Wheeler wondered whether the Schwarzschild singularity signaled where another breakthrough lay hidden in fundamental physics.

The Princeton team began their work on gravitational collapse by following up and extending the work of Oppenheimer and his cohort, an endeavor in which the Princetonians now had a decided advantage. One of the world's first digital computers—MANIAC, which stood for Mathematical Analyzer, Numerical Integrator, and Computer—was available at the nearby Institute for Advanced Study, making the task of calculation easier. One of the group's early results was presented at an international physics conference in Belgium in 1958. Wheeler and his students, B. Kent Harrison and Masami Wakano, figured that a collapsing star would spew out so much light and matter that it would save itself, ultimately settling down as a stable white dwarf or neutron star. Yes, Wheeler told the audience, like Oppenheimer they had seen that a star more massive than about two suns would implode, squeezing its mass to an extreme density. But did it end up disconnecting from the universe? No, answered Wheeler firmly. To save nature from such an absurdity, they had the elementary particles at the center of the star somehow transform into radiation—"electromagnetic, gravitational, or neutrinos, or some combination of the three. . . . A motion picture of a large mass of [nuclear matter] dissolving away under high pressure into free neutrinos presents a fantastic scene," the team reported. This allowed enough mass to escape that the star safely settled down, perhaps as a mere neutron star—not a singularity. There was as yet no physics to fully explain this mechanism, but there was also no physics to prove it absolutely wrong. How gravity acted at the quantum level, in spaces no bigger than an elementary particle, was still a big unknown. It's a condition, the Princeton group noted, "which lies at the untamed frontier between elementary particle physics and general relativity."

Oppenheimer was in the audience and at the end of Wheeler's talk took the floor and politely disagreed. Why count on new physics appearing down the line? "Would not the simplest assumption about the fate of a star of more than the critical mass be this," he asserted, "that it undergoes continued gravitational contraction and ultimately cuts itself off more and more from the rest of the Universe?" Oppenheimer thought the entire matter had already been solved with his 1939 paper. But Wheeler, as yet, was not convinced. "It is very difficult to believe 'gravitational cutoff' is a satisfactory answer to the problem," he replied. Like Eddington before him, Wheeler was hoping to theoretically wipe the singularity off the face of the cosmos.

But with further work on this problem, Wheeler and his students soon learned that a star of great mass would not be stopped from its collapse in the way they had described. Their radiative model simply didn't work. Determined, Wheeler considered the next potential loophole on his list. Perhaps electromagnetic forces come into play. The repulsive forces between particles of similar electric charges might be powerful enough to halt the collapse. But again, their calculations proved that the gravity of all that collapsing matter overwhelmed such electromagnetic forces.

Kip Thorne arrived in 1962 to join Wheeler's burgeoning quest. He went to Princeton specifically to work with Wheeler and clearly remembers, on entering Wheeler's office, being greeted like an esteemed colleague rather than "a green graduate student" looking for a thesis topic. Top on Wheeler's agenda were the many unresolved aspects of gravitational collapse, which they thoroughly discussed. "I emerged, an hour later, a convert," said Thorne.

Theorists in the Soviet Union, most notably Yakov Zel'dovich, were already ahead in this game. In the West, physicists had largely ignored the papers on gravitational collapse until Wheeler came along. "In Western circles," noted Werner Israel, "the work of Oppenheimer and Snyder was a forgotten skeleton in the cupboard . . . dismissed as the wildest speculation." But the Soviets had embraced their papers far earlier. Having those articles on Landau's "golden list" gave the Russians the incentive to pay attention. Landau had also included the Oppenheimer-Snyder results in a well-used textbook he coauthored. If a star was massive enough, the textbook stated in 1951, "a body must tend to contract indefinitely." Soviet physicists didn't doubt Landau's wisdom at all; he was so revered that they took continued gravitational collapse for granted.

Like Wheeler, Zel'dovich had a background in nuclear physics. Working as a lab assistant right out of high school, carrying out impressive research, he learned so much chemistry and physics on his own that he was awarded a doctorate in his early twenties without having attended classes at a university. He went on to become a key member of the teams that built the first Soviet atomic and hydrogen bombs. In fact, his knowledge of astrophysics aided his work on the bomb. Dipping deeply into a book by Landau on gas dynamics, he and his teammates, including Andrei Sakharov, came to recognize that "the physics of stars and the physics of a nuclear explosion have much in common."

Each pioneer, in the West and in the East, placed his own stamp on their "intellectual progeny," as Thorne put it. "Wheeler was a

Yakov Zel'dovich (*American Institute of Physics Emilio Segrè Visual Archives, Physics Today Collection*)

charismatic, inspirational visionary," said Thorne. He'd sometimes provide some general ideas but largely encouraged his students to become independent researchers, offering advice when needed. If their research took time, that was fine with him.

On the other hand, "Zel'dovich was the hard-driving player/coach of a tightly knit team," continued Thorne. Everyone on the team vigorously explored an idea together, trying to keep up with Zel'dovich's blazing intellectual pace. In his camp, one and all got credit. Both Wheeler and Zel'dovich handed down those separate styles and approaches to the next generation of general relativists, who carried black-hole research into its golden age.

A turning point for both sides arrived in the mid-1960s when physicists were at last able to successfully simulate the implosion of a stellar core on its death, using the same advanced computers and same mathematical techniques that allowed physicists to design nuclear weapons. Thorne recalled Wheeler rushing into a relativity class one day with the news of the latest results from these simulations, carried out at the Livermore National Laboratory in California by the endeavor's chief gurus Sterling Colgate and Richard White. Wheeler often traveled to Livermore to keep tabs on their work. "When the mass of the star was much larger than the 2-Suns maximum for a neutron star, the implosion—despite its pressure, nuclear reactions, shock waves, heat, and radiation—produced a black hole. And the black hole's birth was remarkably similar to the highly idealized one computed nearly twenty-five years earlier by Oppenheimer and Snyder," said Thorne. If the stellar core was weighty enough, it turned out that *nothing*—no other force in the entire universe—could stop gravity from creating a black hole.

In the Soviet Union Zel'dovich, too, saw that his know-how in bomb design could apply to simulating a star collapsing. Both Cold War adversaries knew this, but neither dared discuss their bomb work with each other. "I had many discussions with [Zel'dovich] and shared a sleeping car compartment with him one time from Warsaw to Moscow, and we never talk[ed] about that subject," recalled Wheeler. "But Zel'dovich one day was writing on the board a formula for the explosion of a star. He gave me a wink, and I winked back. He and I knew it came from another context." And those bomb-related calculations, carried out independently in the Soviet Union, arrived at the very same answer as in the West. Black holes were inevitable.

With such evidence in hand, Wheeler utterly reversed his earlier, negative opinion concerning the singularities of completely collapsed stars. Formerly determined to get rid of black holes somehow, he now became their greatest champion. But it was more than his own theoretical deliberations and the computer simulations that finally convinced him. Also crucial was a new way to look at a black hole. If you were viewing a star collapsing from very far away, you would never see the star fully shrivel down to nothingness. Due to time dilation effects, you would only get to see the star's surface "freeze" into place just as it reached the critical circumference—its event horizon. Why is that so? It's because, as Einstein pointed out very early in his work on general relativity, time slows down in a gravitational field, and as the star gets denser and denser in its collapse, it takes longer and longer for the star's photons to escape—and it's those photons that allow us to see what's happening. By the time the star reduces to the size of its event horizon, then it takes an *infinite* amount of time for us to see any further progression. Time stops in its tracks. It's like a movie projector going slower and slower, until it halts and lingers on just one image. That's why Soviet scientists gave such a collapsing object the name "frozen star."

But that doesn't mean the star is actually freezing into place. In *its* frame of reference (not ours from afar), total oblivion is swift. If you were magically shifted onto the star itself, falling inward with the collapse, you'd pass right through the horizon without hesitation. The two different reference frames depict different outcomes simply because they do not share the same space and time. "You cannot appreciate how difficult it was for the human mind to understand how both viewpoints can be true simultaneously," Russian physicist Evgeny Lifshitz told Kip Thorne.

But in 1958 David Finkelstein, then a young, little-known physicist at Stevens Institute of Technology in New Jersey, developed a new reference frame to handle these different viewpoints concurrently. A new perspective, if you will. It allowed physicists to picture how a collapsing star appears like a frozen star to us from afar yet still fully implodes from the standpoint of the hole. Plasma physicist Martin

Kruskal had actually arrived at a similar result earlier. In the mid-1950s he had joined a small group of Princeton colleagues who wanted to learn general relativity on their own, and during that time Kruskal developed an even more extensive framework than Finkelstein later demonstrated. When Wheeler showed no interest in the new derivation, Kruskal just put it aside. A couple years later, though, Wheeler finally realized that his indifference had been an egregious oversight and soon wrote a paper on the coordinates under Kruskal's name (to Kruskal's surprise). It was published in 1960.

In the end, both Finkelstein and Kruskal made it easier for theorists to visualize—all at once—every strange, relativistic effect going on both from our vantage point here on Earth and right at a black hole's event horizon far away in space. That made the physics far more comprehensible for Wheeler's team. And it also broke the logjam in tackling relativistic problems once thought impossible to solve. "The field of gravitation was then almost completely dominated by Newton's theory," notes historian Jean Eisenstaedt, "and Kruskal's interpretation came as a bombshell in the small sleepy relativist village."

By 1962 Charles Misner, then at Princeton and working with Wheeler, recruited undergraduate David Beckedorff to take these new mathematical tools and essentially redo Oppenheimer and Snyder's work for his senior thesis. According to Misner, Beckedorff's treatise was the first description of the space outside an imploding star, showing how the matter falling inward crossed the event horizon. "Even if you sent a suicidal spacecraft at the speed of light, to try to catch up to that collapsing star, you'd never make it," explains Misner. You'd have to go faster than the speed of light. This image wasn't available in the Oppenheimer-Snyder paper. Beckedorff's solution, worked out under Misner's guidance, introduced an entirely new way to look at a black hole.

Before this physicists and relativists working on gravitational collapse were concerned only with the stellar matter: What was happening to the stellar material, what was its final state? "But it's gone," says Misner. "What's left is the black hole. Previously people focused on the fate of the star, but now we were seeing that something had formed. It's still there, and it can do things. It's not just the grave-yard of the star." That, too, was a turning point for Wheeler. This new perspective allowed him and others to recognize the black hole as a real object, even though its mass is now hidden behind the event horizon.

And about that term *event horizon*—it was physicist Wolfgang Rindler, then at Cornell University, who first used the phrase in 1956. Only in his case, he was applying it to cosmological models of the universe. On one side of an event horizon, events can be seen by us; on the other side, said Rindler, they are "forever outside [our] possible powers of observation." In the cosmological case, the celestial objects in our expanding universe have flown past the borders of the visible universe. At that far point, their light waves can never catch up to us as the universe continues to expand. It turns out this was also the perfect definition to describe Schwarzschild's point of no return. Once an object is inside the event horizon, it can never again be seen from the outside. So, by the early 1960s, astrophysicists began using the term as well when talking about the outer boundary of a gravita-tionally collapsed star.

Exhilarated by their successes, physicists in the Soviet Union, the United States, Great Britain, and continental Europe began to exam-ine the characteristics of a black hole in more detail. They looked at each and every property a black hole could possibly have. What happens to the collapsing star's magnetic fields, for example, as the event horizon emerges? They get severed from the dying star and snap off like rubber bands. From the outside, the lone black hole ends up having no magnetic field at all.

What if the collapsing star is deformed? Nothing in nature is perfect; maybe even a minor bump or bulge on a star would halt its collapse. Simulations up to this point usually had a perfectly spherical star, which perhaps falsely caused the virtual star to collapse uniformly within the computer calculations to an exact point. And, for a while, it appeared that would be physics' savior in preventing the formation of singularities. In 1961 two Russians, Evgeny Lifshitz and Isaak Khalatnikov, seemed to have proven that irregularities did make a difference. In their simulations, they started with a lumpy star and found that some parts of this star would collapse faster than others and so experience a rebound at the center, preventing a singularity from ever forming. They went so far as concluding that singularities would never be created in a real universe. But that turned out not to be the case. Within several years they found a mistake in their calculations and, once corrected, arrived at the exact opposite conclusion. No matter what the stellar matter looked like at the start, the collapse is not halted and the end product—the black hole's horizon—is blandly uniform.

Case by case led the relativists to one, undeniable conclusion. No matter what a star looks like before its gravitational collapse, all its distinguishing features vanish—leaving behind only *three* pieces of information. The only properties that remain are the former star's mass, its spin, and its electric charge (though that charge would likely be neutralized after attracting equal and opposite charges from its environment). As John Wheeler liked to say, "A black hole has no hair," no defining characteristic that makes it look different from any other black hole. "There is no way to tell from outside whether a black hole was created using neutrinos, or electrons and protons, or old grand pianos." Or if it had been yellow, wrinkled, or polka-dotted. A black hole is a black hole is a black hole. Every distinctive feature of a star disappears behind its inscrutable event horizon. The gravitationally collapsed object should not be thought of as a frozen star at all. Instead it should be thought of as a soap bubble–like pure gravitational field—its only properties being mass, angular momentum, and electrical charge. The singularity itself would never be seen, forever cloaked behind the event horizon.

That an object so cosmic could appear so basic—described by a mere three numbers—was simply flabbergasting to everyone. It meant that each black hole is as elementary an entity as an electron or a quark. Here was the beauty and simplicity that Chandrasekhar referred to in his Nobel Prize lecture.

Armies of students worked on all the nuances of these problems for years, in order to nail down each and every feature of a black hole. There was always the chance that something might turn up to prevent a black hole from forming after all. But nothing turned up. "The matter of the core pours torrentially inward from all directions like a thousand Niagara Falls on its way down from the original [stellar] dimensions to ever smaller sizes," wrote Wheeler in 1968. "In less than a tenth of a second . . . the collapse goes to completion, and little if anything is seen."

Roger Penrose had already provided powerful ammunition to back up Wheeler's statement. Penrose was a member of a highly influential group of British theorists who were devising a number of brilliant topological and geometrical tools to answer key questions about the physics of black holes. Trained in mathematics, not physics, Penrose first became interested in relativistic singularities after hearing Finkelstein lecture in the late 1950s in London. "When I got back to Cambridge, knowing very little about general relativity," said Penrose, "I started to try and prove that singularities were inevitable. It seemed to me that maybe this was a general feature." Yet, at the same time, they also seemed to him a bit "ridiculous and mysterious," he noted. After intermittently working on the problem over the ensuing years, he at last published a theorem in *Physical Review Letters* in 1965 that has been called by some as "the most influential development in general relativity in the 50 years since Einstein founded the theory." Four years before Lifshitz and Khalatnikov discovered the mistake in their calculations, Penrose proved, in less than three pages, that full gravitational collapse and singularities go hand in hand. It took a while to convince everyone because he used a mathematical approach not familiar to most physicists. But the bottom line was unmistakable: you couldn't have total gravitational collapse without a singularity turning up at the end. "Deviations from spherical symmetry," reported Penrose, "cannot prevent space-time singularities from arising." (That is, as long as quantum mechanics is not taken into account, which black-hole theorists were not yet doing; more on that in chapter 12.)

Wheeler knew, Penrose knew—and current physics still contends—that given enough mass total collapse is inescapable. "The core like the Cheshire cat fades from view. One leaves behind only its grin, the other, only its gravitational attraction," Wheeler has said. Everything we know about the strength and stability of matter against the unyielding force of gravity leads to that unavoidable end. The mass disappears from our view; only its gravitational attraction remains behind to affect us.

Wheeler tried to impart the insanity of the concept. Ride along on the collapsing ball of matter, he suggested, and its density would go up faster and faster until, in less than a second, it rose to infinity.

"With this prediction of an infinite density," wrote Wheeler, "classical theory has come to the end of the road. A prediction that is infinity is not a prediction. Something has gone wrong. . . . Infinity is a signal that an important physical effect has been left out of account."

And there is something missing. Wheeler pointed out that answers will likely arrive with the successful merging of general relativity with quantum theory. Today we call it superstring theory or loop quantum gravity, the latest attempts to join the macrocosm ruled by gravity with the microcosm controlled by quantum forces. Quantum-gravity theorists don't have as yet a definitive solution, but they are confident that something else happens inside the black hole, that quantum effects prevent a singularity from forming there.

In the early 1970s, one final escape hatch to prevent the creation of a black hole remained: pulsations. Carrying out computer simulations, researchers saw that a black hole can also vibrate—in a sense, ring like a bell if it's disturbed. Could these pulsations get unstable and, by extracting energy from the hole, get stronger and stronger— so strong that the hole is violently torn apart? The answer in the end was an unequivocal no. The extra energy simply radiates away from the hole as gravitational waves, ripples in the very fabric of space-time. The black hole remains intact.

Tackling such relativistic problems was a brave, if not foolish, choice for a rising physics student at the time. Few then believed that a star underwent any kind of collapse. Thorne remembered being warned in the early 1960s that general relativity had "little relevance for the real Universe. . . . One should look elsewhere for interesting physics challenges." Fortunately, Thorne ignored such skeptics. Those naysayers soon learned that general relativity was crucially needed in astrophysics—and in a big way. That's because, while Wheeler, Zel'dovich, and others were grappling with the theory of gravitational collapsed objects and the revival of general relativity, astronomy was undergoing its own parallel revolution. It's when observers started to gather an array of celestial radiations other than visible light, leading to some unexpected discoveries that cried out for explanation.

8

It Was the Weirdest Spectrum I'd Ever Seen

A new astronomy for the twentieth century arose in a very unusual spot. It took place amid central New Jersey's potato fields. In the 1930s, Karl Jansky set up a unique radio receiver near the rural town of Holmdel and, in doing so, became the first person to snatch astronomy away from its dependence on the optical spectrum, beyond the narrow band of electromagnetic radiation visible to the human eye. His first, provisional step ultimately led to a new and golden age of astronomy that thrives to this day. But, as is often the case in astronomical history, Jansky began his investigations for a totally different reason.

In 1928, fresh out of college with a degree in physics and newly hired by Bell Telephone Laboratories, the twenty-two-year-old was assigned to investigate long-radio-wave static that was disrupting transatlantic radio-telephone communications. To track down the sources, he eventually built a steerable antenna—a spindly network of brass pipes hung over a wooden frame that rolled around, with a motorized push, on Model-T Ford wheels placed on a concrete track. It was known around the lab as "Jansky's merry-go-round."

Setting up his antenna near Bell's Holmdel station, Jansky soon learned that thunderstorms were a major cause of the disruptive clicks and pops during a radio phone call. But there was a steady yet weaker hiss that he also kept receiving. After a year of detective work, Jansky

Karl Jansky with the "merry-go-round," his historic antenna that discovered radio waves emanating from the center of the Milky Way galaxy, initiating the field of radio astronomy. (*Reprinted with permission of Alcatel-Lucent USA Inc.*)

in early 1933 at last established that the disruptive 20-megahertz static (a frequency between the United States AM and FM bands) didn't originate in the Earth's atmosphere or on the Sun or from anywhere within our solar system. To his surprise, he saw that it was coming from the direction of the Sagittarius constellation, where the center of our home galaxy, the Milky Way, is located. Jansky affectionately dubbed the signal his "star noise." For Jansky it hinted at processes going on in the galactic core, some twenty-seven thousand light-years distant, that visible light rays emanating from that region did not reveal. For unlike visible light, radio waves can cut through the intervening celestial gas and dust, in the manner of a radar signal passing through a fog.

Jansky's unexpected discovery made front-page headlines in the *New York Times* on 5 May 1933, with readers being reassured that the galactic radio waves were not the "result of some form of intelligence striving for intra-galactic communication." Ten days later NBC's public affairs–oriented Blue Network broadcast the signal across the United States for the radio audience to hear. One reporter remarked that it "sounded like steam escaping from a radiator."

By 1935, Jansky speculated that the cosmic static was coming either from the huge number of stars in that region or from "some sort of thermal agitation of charged particles," which was closer to the truth. Years later, astronomers confirmed that the noise was being emitted by violent streams of electrons spiraling about in the magnetic fields of our galaxy. Just as an electric current, oscillating back and forth within an earthbound broadcast antenna, releases waves of radio energy into the air, so these energetic particles broadcast radio waves out into the cosmos, whose wavelengths are far longer than visible light. And Jansky was the first to detect them. He was Earth's first eavesdropper on the universe.

Despite the worldwide publicity, however, few astronomers then appreciated Jansky's new ear on the universe. Most were more comfortable with lenses and mirrors than with radio equipment. "The world of decibels and superheterodyne receivers . . . was far too removed from that of binary star orbits and stellar evolution for a connection to be forged," explains science historian Woodruff Sullivan. And Bell Labs did no follow-up, since astronomy wasn't its business at the time. The company put Jansky to work on more commercial problems. But one particular person was inspired by the Bell Labs employee's innovation. An Illinois radio engineer and avid ham-radio operator, Grote Reber, erected a massive steel saucer—a thirty-foot-wide dish antenna—in his backyard and extended Jansky's work. He showed that celestial radio waves were most intense along the plane of the Milky Way. In 1940 he sent his results to the *Astrophysical Journal,* which turned out to be the first paper on radio astronomy the publication had ever received. Only the intervention of a farsighted editor kept it from being rejected. Four years later Reber produced the first map of the entire "radio sky." Along with the strong peak at the Milky Way's center, there were secondary peaks in the directions of the Cygnus and Cassiopeia constellations.

Over this time World War II had intervened, essentially slowing any progress, but afterward the field of radio astronomy took off. The war itself was actually one of the reasons. Dozens of young physicists and engineers in Europe, Australia, and the United States had been introduced to the esoteric art of radio science while working on the development of radar during the conflict. After the war they were eager to apply their newfound skills to follow up on the work of radio astronomy's two pioneers. They wanted to pinpoint the celestial objects putting out those mysterious radio signals. To this vanguard, the radio sky was a blank page just waiting to be filled in. What happened next has been called "the most eventful era in the history of astronomy since the time of Galileo."

Radio telescopes began cropping up around the globe, with England and Australia dominating the field at first. They found the nebulous remnants of ancient supernovae emitting loud radio squeals. And Cygnus A, one of the "brightest" objects in the radio sky, turned out to be a strange-looking galaxy located about six hundred million light-years away. Similar "radio galaxies" were found all over the heavens. By developing techniques to combine the signals from radio telescopes separated by a mile or more, which together then acted as one large telescope, they obtained enough resolution to see that the radio signals from these peculiar galaxies were emanating from giant lobes of gas, jutting out for a few hundred thousand light-years from the galaxy like the wings on a plane. How did they possibly originate?

The solution involved thinking about the universe in a whole new way. No longer was it just stars and galaxies floating in space but also particles like electrons racing within the electromagnetic fields filling interstellar and intergalactic space. It was these electrons that were releasing the radio waves as they spiraled about the magnetic

Two giant radio lobes, each about six hundred thousand light-years across, encase the giant elliptical galaxy, Fornax A (seen in the center), situated some sixty million light-years from Earth. (*Courtesy of NRAO/AUI and J. M. Uson*)

field lines. By 1958 astrophysicist Geoffrey Burbidge figured that those gigantic lobes surrounding radio galaxies held as much magnetic and kinetic energy as if the matter of ten million suns had been *completely* converted into pure energy, according to $E = mc^2$. Optical telescopes had been blind to this activity, making astronomers believe for centuries that the universe was fairly serene. But by reaching out into broader regions of the spectrum, astronomers now knew that something *big* was going on in distant space that traditional energy sources couldn't account for. The universe was jam-packed with action.

Chemical power, like that from dynamite, was far too weak; even nuclear energy appeared dicey. "Nuclear fuel's efficiency for mass-to-energy conversion is roughly 1 percent," Kip Thorne once estimated. That means an active galaxy would need one billion solar masses of nuclear fuel to energize its radio-emitting lobes. Possible, but not likely. Energy drawn from matter annihilating antimatter was briefly considered but also discarded as a possible source. The universe just didn't seem to harbor enough antimatter.

The mystery got even stranger. By the late 1950s, spurred on by these discoveries and not wanting to miss the boat, the United States built its own state-of-the-art radio observatories. One of them, a complex situated in California's Owens Valley and run by Caltech, was able to narrow down the location of a radio source labeled 3C 48, for being the forty-eighth radio object in the Third Cambridge Catalogue of radio sources. Astronomer Allan Sandage soon used the grand 200-inch Hale telescope atop California's Palomar Mountain to see what visible object might be situated at that spot. After carrying out a ninety-minute exposure and expecting to see another galaxy there within the Triangulum constellation, he instead found a pinpoint of light, a real surprise. To the eye, it was yellow in color, but also unusually bright in the ultraviolet region of the spectrum. At first, everyone just assumed it was a star in our own galaxy, making it the first known "radio star." But there was a catch: "I took a spectrum . . . ," said Sandage, "and it was the weirdest spectrum I'd ever seen."

Over the following two years, a handful of similar objects were discovered. On first look they appeared to be simply faint stars within the Milky Way, just like 3C 48. But again, after viewing the light waves emanating from these radio stars more closely, optical astronomers found that they all displayed spectral features unlike any star ever observed. The spectra of these stars didn't match any known chemical element. Could there be other chemicals out there, as yet undiscovered? It was like riding down a familiar turnpike and finding that all the road signs were written in gibberish. Astronomers couldn't even find evidence that hydrogen—the main component of all stars—was present. Yet, everyone kept assuming they were stars because, well, they *looked* like stars through an optical telescope. For one, they flickered. If this strange object were a far-off galaxy, it was considered "utterly ludicrous" to believe that some one hundred billion stars could turn their brightness on and off in sync so swiftly. Not until February 1963 was the identity of these peculiar radio beacons finally unmasked.

On the fifth day of that month, thirty-three-year-old Maarten Schmidt, who had arrived a few years earlier at Caltech from the Netherlands, was sitting at his desk attempting to write an article on the radio star known as 3C 273 for the British journal *Nature*. Australian radio astronomers had just gone to extraordinary lengths to view this source. They cut down trees and tipped a weighty radio dish beyond its safety limits to better pinpoint the position of this radio star caught low on the horizon. With the improved coordinates, Schmidt was able to use the Palomar telescope to find the star in visible light and obtain an optical spectrum. With the spectrum spread before him, Schmidt came at last to recognize a familiar pattern of spectral lines that had eluded him for weeks. The pattern resembled the specific wavelengths of light typically emitted by simple hydrogen when energized—but they were in the wrong place! That's why hydrogen had appeared to be missing. The hydrogen lines were there, but shifted *waaaay* over, toward the red end of the spectrum. That meant this starlike object was moving *away* from us at a tremendous speed. Just as the pitch of an ambulance siren gets lower as it races away from us, a light wave is stretched to longer lengths (gets "redder") when its source recedes—a type of Doppler shift. Consequently, this "redshift" lets astronomers gauge both how fast a celestial object is moving and also its distance.

In this way, Schmidt swiftly grasped what that redshift meant. It turns out that 3C 273 was not an unusual star situated within the Milky Way but rather a bizarre object located some two billion light-years away (one of the farthest cosmic distances ever recorded at that time). It was rushing away from us at nearly thirty thousand miles (forty-eight thousand kilometers) per second, carried outward with the swift expansion of the universe. Schmidt knew that only an incredibly bright source could be visible from such a distance; he figured that 3C 273 was radiating the power of trillions of stars and suspected it was the brilliant and very disturbed nucleus of a distant galaxy. This galaxy appeared starlike only because it was so far away.

With that revelation, all fell into place. The spectra of other mystifying radio stars were quickly deciphered. These blue, extragalactic specks were soon referred to as quasi-stellar radio sources (QSRS) or quasi-stellar objects (QSO) by the California astronomers. Before long, they were simply called quasars, a term at first scorned by old-school astronomers. Not until 1970 did Chandrasekhar, by then the *Astrophysical Journal*'s editor-in-chief, permit its official use and that was only after Schmidt convinced him that the name could no longer be ignored. "*The Astrophysical Journal* has up till now not recognized the term 'quasar,'" Chandra wrote in a footnote to one of Schmidt's papers, "and it regrets that it must now concede."

The quasar 3C 273 is now considered relatively close to us, as quasars go. Its distance is small potatoes compared to those of later finds.

The first known quasar, 3C 273, as imaged by the Hubble Space Telescope's Wide Field Planetary Camera 2. The diffraction spikes demonstrate the quasar is truly a point-source of light, like a star, and hence very compact. (*Courtesy of NASA/Space Telescope Science Institute*)

Over the past five decades, astronomers have now identified quasars out to a distance of some thirteen billion light-years, which means they were bright and active less than a billion years after the Big Bang. The fact that earthbound observers are able to see such quasars across the vastness of the universe means that these objects are the most powerful denizens of the heavens.

But what could possibly be the source of such monstrous energy, everyone asked upon the quasar's discovery? "The insult was not that they radiate so much energy," said Schmidt, "but that this energy was coming from a region probably no more than a light-week across." Astronomers came to know this by seeing the quasars dim and brighten over a matter of weeks or days. In the case of 3C 273, they checked old photographic plates of the thirteenth magnitude object (roughly four hundred thousand times fainter than the star Sirius), going back some seventy years. In one picture it was faint, a month later it was brighter. Such relatively swift fluctuations meant that the quasar's power source was small, perhaps less than the diameter of our solar system. That's because any quick luminosity changes in a vastly larger object would get lost in the noise. Yet from such a cosmically tiny region spewed the energy of billions of suns. Tapping into such a cosmic dynamo for just one second would power the world for a billion billion years. What cosmic process could conceivably generate such energies?

Suddenly any idea, no matter how far-fetched, was given consideration. "The discovery of quasars," noted Schmidt, had "a profound impact on the conduct of those practicing astronomy. Before the 1960s, there was much authoritarianism in the field. New ideas expressed at meetings would be instantly judged by senior astronomers and rejected if too far out. . . . [But now] an attitude . . . evolved where even outlandish ideas in astronomy [were] taken seriously."

Fred Hoyle and William Fowler, for example, dared to bring up general relativity, for so long ignored. A month before Schmidt even announced his identification of 3C 273 in *Nature,* Hoyle and Fowler published an article in the same journal pointing to gravity as a possible cosmic engine, in this case for the many active radio galaxies already found closer in. They imagined that up to one hundred million solar masses had accumulated in the center of these galaxies, behaving as one giant star. A sudden contraction of this mass "to the relativity limit"—that is, a catastrophic gravitational collapse—would then release the immense energies displayed by these galaxies. This extended an idea first published two years earlier by the Soviet physicist Vitaly L. Ginzburg, who had studied under Landau.

Schmidt's discovery of quasars, coupled with Hoyle and Fowler's intriguing theory, swiftly reverberated throughout the physics and astronomy communities. A special conference was quickly organized by a group of top relativists to bring astronomers, theorists, and physicists together to discuss the myriad questions on everyone's mind concerning quasars. Funders for this large gathering included NASA, the US Navy, and the US Air Force. (Some in the military were interested because of that short-lived flurry of excitement in certain quarters, encouraged by Roger Babson's enthusiasm, that the study of general relativity might lead to antigravity devices.) "For more than ten years," said the conference's invitation, "the nature of the strong extragalactic radio sources has been one of the most fascinating problems of modern astronomy. . . . [The enormous] energy requirement has so far ruled out nearly all of the explanations and theories put forward to explain such extraordinary events. . . . It emerges that the many-sided and basic aspects of the question of gravitational collapse make it imperative to bring experts from many fields together for a thorough discussion."

J. Robert Oppenheimer was among those invited. It looked as if his 1939 paper, neglected for nearly a quarter of a century, was finally going to get its day in the Sun. The science world was electrified. "I look forward to seeing you at the Dallas conference," Penrose wrote Wheeler. "The subject is certainly a most intriguing and baffling one."

9

Why Don't You Call It a Black Hole?

The conference might never have happened were it not for strong martinis, coupled with a hot, lackluster Texas summer. The noted mathematician Ivor Robinson had just moved to Dallas at the start of 1963 to head the newly formed relativity group at the Southwest Center for Advanced Studies (what later became the University of Texas, Dallas), and he was bored. As one observer put it, he was pining "for people who would recognize . . . a null bivector when they saw one." So, over the long Fourth of July weekend, he invited a number of friends to visit his locale, then far from the usual watering holes in the field of general relativity.

While everyone was lazily fanning themselves as they sat around a suburban Dallas swimming pool on July 6, drinks in hand, the center's chief scientific officer, physicist Lauriston Marshall, suggested that a little conference, perhaps with some twenty-five participants, might be just the thing to put their new institute on the map. "Give a little spice to life," he said. Robinson, along with relativists Alfred Schild and Engelbert Schücking, visiting from the University of Texas at Austin, jumped at the idea.

While the three mulled over potential topics in the ensuing days, Schücking happened to mention the newly discovered quasars. "Nobody knows quite what they are," he pointed out. "Why don't we hold a conference on the subject?" Everyone agreed, but they recognized that such a big topic required a far bigger platform than they originally planned. So, the initial idea to hold an intimate workshop was soon "Texanized . . . [into] a big bash in Dallas," as Schücking put it. The Dallas money, coming from the city's establishment to help build a Princeton in Texas, was "particularly valuable," added Schücking, "since it could be spent on liquor, while the Lone Star State money from Austin could be used only for more sober expenses."

But what to name the conference? Here was a small relativity group holding a conference on what was largely an astronomical subject. "We fixed that," said Schücking. He, along with Schild and Robinson, invented the name for an entirely new field. The conference title, they decided, would be "The Texas Symposium on Relativistic Astrophysics," and they invited nearly everyone they could think of with a connection to this fresh discipline. "Relativity was the sleeping beauty and quasars the prince that woke her up," says German science historian Jürgen Renn. It was the moment when many top physicists first learned that general relativity might actually be significant in the physical world.

The meeting took place in December 1963, right before the Christmas holidays, at a hotel in downtown Dallas, just blocks from the street where President John F. Kennedy had been assassinated three weeks before. Conference organizers had been urged to cancel the event because of the tragedy, but they decided to stay the course. Texas governor John Connally, also injured in that terrible episode, welcomed the conferees during the opening address with his arm in a cast.

Three hundred scientists attended from around the world, with Oppenheimer chairing the first session. Schücking recalled how, just minutes before this session began, Oppenheimer "asked us to synchronize our watches. It was as if we were going to have another

Alamogordo." Also present was Karl Schwarzschild's son Martin, who had grown up to become a Princeton University astronomer. All these relativists, astronomers, and astrophysicists felt a palpable excitement in the air. As one participant put it, "Many of those who attended felt that they [were] at a historic occasion where new ideas of paramount importance were presented that may profoundly influence all future thinking in the field."

It was the great convergence. At last general relativity and astrophysics were being directly linked. By the time of the conference, astronomers had identified nine quasars. Had Oppenheimer's gravitationally collapsed objects at last been sighted? As Cornell University astrophysicist Thomas Gold wittily suggested in an after-dinner talk at the meeting, relativists were now more than "magnificent cultural ornaments but might actually be useful to science! Everyone is pleased: the relativists who feel they are being appreciated, who are suddenly experts in a field they hardly knew existed; the astrophysicists for having enlarged their domain, their empire, by the annexation of another subject—general relativity. . . . So let us all hope that it is right. What a shame it would be if we had to go and dismiss all the relativists again."

The invitation to the conference outlined a clear and concise agenda. "Among the problems raised are the following," said the notice:

(a) The astronomers observed some unusual objects connected with radio sources. Are these the debris of a gravitational implosion?

(b) By what machinery is gravitational energy converted into radio waves?

(c) Does gravitational collapse lead . . . to indefinite contraction and a singularity in space-time?

(d) If so, how must we change our theoretical assumptions in
order to avoid this catastrophe?

That last statement was particularly interesting. It suggests that
physicists were still hoping to sweep the ugly singularity under the
rug. Despite the work of Oppenheimer and Snyder, Wheeler and
Zel'dovich, gravitational collapse remained a hard pill to swallow.
Some attendees didn't even know the possibility existed until they
attended the conference. It was the first they heard of such a stellar
outcome.

One question was on the top of everyone's minds: What was the
energy source for those objects putting out such intense radio and
optical radiation? The object 3C 273 was emitting the energy of a trillion suns. How long had this been going on? How long would it last?
Scientists already knew that nuclear reactions were simply too ineffi-
cient to produce such a flood of continuous energy. That was the
reason the spotlight was on gravity—gravitational collapse, to be
exact. As matter accelerates in its fall toward a black hole, immense
energy is released, far more than if that matter was used as fuel in
nuclear burning.

On the first morning of the conference, Fred Hoyle and William
Fowler discussed their idea of a giant body of matter contracting. In
its outer region, thousands of pockets of condensed matter, each some
one hundred solar masses, go through nuclear burning, not over billions of years, as in the case of our Sun, but in a swift "whoosh," over
the matter of a week. At the same time, in the innermost region of this
gargantuan body of matter, contraction continues. A "superstar,"
weighing some one hundred million solar masses or more, emits a
tremendous blast of energy as it catastrophically shrinks down to a
point. How did such a mammoth star get there in the first place? "For

From left to right, Fred Hoyle, Ivor Robinson, Engelbert Schücking, Alfred Schild, and Edwin Salpeter at the 1963 Texas Symposium on Relativistic Astrophysics. (*American Institute of Physics Emilio Segrè Visual Archives, E. E. Salpeter Collection*)

the moment," wrote Hoyle and Fowler in their published remarks, "we ignore the question of how such an object might be formed—the observational evidence would seem to give strong support to the postulate of the existence of massive objects, and it is therefore reasonable to inquire into their properties without further ado. We turn a blind eye, a deaf ear, and a cold shoulder to written, oral, and implied criticism, respectively." Hoyle and Fowler went on to suggest that a kind of "antigravity" field (something, of course, not seen in physics) kept the superstar from total collapse to a singularity; pushing outward, they wrote, the field causes the crushed star to bounce back, releasing the observed radiation. These oscillations would continue but slowly subside over time, until the object shrinks entirely behind its event horizon and vanishes from sight. Maarten Schmidt was right; astronomers had entered an era where, as the noted music composer Cole Porter put it, "anything goes."

Some held open the possibility that the extremely intense gravitational field of this collapsed superstar—not light waves being stretched over long distances in an expanding universe—generated the quasar's tremendous redshift. This meant that the quasar could be massive yet small, and relatively close by. But it was soon settled that this couldn't be the case; if it were true, the motions of stars in our Milky Way, situated near such a massive object, would be greatly altered by its strong gravity field. Such deviations were not observed at all in our local galactic neighborhood. By some calculations, if 3C 273 were such a galactic star, it couldn't be more than a third of a light-year distant from us, practically inside our solar system, which would have greatly affected planetary motions. And if such a star were receding from us at thirty thousand miles (forty-eight thousand kilometers) per second, our galaxy couldn't gravitationally hold onto it for very long.

Other mechanisms were also considered. Perhaps the quasar phenomenon arose when matter and antimatter annihilated one another in a galaxy's center? When a bit of matter meets up with its antimatter counterpoint, the pair obliterate each other, leaving nothing behind but a burst of pure radiation. But how could these disparate materials be kept apart for so long before that?

Conference participants excitedly argued over these various ideas, one after the other. Could it be that a quasar was a massive chorus of supernovae exploding all at once? Not really. To produce such energy, you'd need a hundred million going off. Why (and how) could so many stars be exploding simultaneously? Moreover, each million or more stars would have to be stuffed into a volume only a few light-years wide. Is that even possible?

A singular and sudden gravitational collapse, the kind that Hoyle and Fowler discussed, had a problem as well. Physicist Freeman Dyson emphasized this point. A collapse would generate lots of energy, he pointed out, but only for a short time—a day at most. Yet quasars shine on and on and on. Gravitational collapse is over quickly, but quasars blaze fiercely for a million years or more.

The Soviet physicists Yakov Zel'dovich and Igor Novikov would soon point out that great energies can be released as nearby dust and gas is drawn toward a massive collapsed object, accumulating into a disk that surrounds the object. This circulating matter radiates and glows for many years as it spirals in, they said, until it ultimately reaches the point of no return and disappears behind the curtain of the event horizon. But such an idea didn't come up at any open session at the conference. (The Soviets were not given permission to travel to Texas.) In the end, no single scheme came out on top by the conference's last day. Rather than counting on new physics to explain a quasar's energy, many held out for more ordinary astrophysical processes, something like gas clouds raining down on a central cluster of stars. Energy would be gained as the gas plunged in toward the galactic nucleus, possibly converted to an explosive form due to shocks and collisions along the way.

It would take a few years for all the varied hypotheses, both bizarre and mundane, to settle down—and bizarre won. It's now generally accepted that a quasar's power source *is* a supermassive black hole, spewing energy as it feeds on an accretion disk of matter swirling around it, just as Zel'dovich and Novikov early on surmised and

Cornell University physicist Edwin Salpeter also independently suggested in 1964. (More on this process in chapter 11.)

Today the First Texas Symposium is also remembered for a brief talk that heralded a major breakthrough in black-hole physics. It came from an up-and-coming relativist named Roy Kerr, but at the time the astrophysicists in the audience hardly noticed. Kerr's presentation wasn't even mentioned at the end of the conference, when three participants presented a summary of the meeting's highlights. But, as you'll see, that gaffe was eventually amended.

While the early 1960s were watershed years for astronomy, what with the discovery of quasars, it was also a time of innovation for relativity, which was on the brink of entering its golden age. For decades, remarked physicist George Gamow, general relativity stood "in majestic isolation, a Taj Mahal of science, having little if anything to do with the rapid developments in other branches of physics." But now with advances in instrumentation, some spurred by the needs of World War II, experimental physicists started rechecking Einstein's predictions with exquisite precision. More than that, they also began carrying out new experiments. "A new and very able younger generation has come on the scene," wrote Wheeler to a colleague at MIT. "Improvement in experimental techniques and a new aggressiveness on the part of the experimenters have broken the subject out of the strait jacket of the old traditional . . . tests of general relativity."

In 1960, for example, Robert Pound and Glen Rebka finally measured the "gravitational redshift," another effect long predicted by Einstein but one that had been highly elusive due to the great precision required. It took four decades of waiting, but Einstein's third prediction of gravity's behavior (along with light bending and orbit shifting) was finally proven. Simply put, a light wave will stretch out (get longer and hence "redder") as it flees from a strong gravitational field. Pound and Rebka measured this very effect on the campus of Harvard University as they directed gamma rays, ejected by radioactive atoms, up a tower several stories high within the physics building. By the time the radiation traveled some seventy-four feet (twenty-two meters) to reach the top, the gamma rays had lengthened just a smidgen—by an amount that matched Einstein's expectation.

It's because of the gravitational redshift that clocks run slower on Earth than in space. You can think of the light waves as springs—coils that get stretched as they climb out the Earth's gravitational well. And as the waves get longer, their frequency—the number of waves passing by us each second—is reduced. If the frequency of those gamma rays were being used as a clock, then the "tick of the clock" is slowed down by Earth's gravitational field. We don't notice this change ourselves, since the very atoms in our body are slowed down as well. We recognize the effect only by comparison. A clock freely floating in space feels no such gravitational effect and so runs faster by comparison. The high-stability clocks aboard the Global Positioning System (GPS) satellites, perched high above the Earth, run a bit faster. As a result, periodic corrections for general relativity must be programmed in to make sure the navigation of our cars, boats, and planes down here on Earth doesn't go awry (perhaps the first time that general relativity was needed to help us in our everyday lives).

How much a clock slows down depends on the strength of the gravitational field. If a person could miraculously survive on the surface of a neutron star, where gravity is a trillion times stronger than Earth's, they would age noticeably slower than a person more loosely grounded on terra firma. As a decade passes by on Earth, the Neutronian would experience about eight years of time going by.

What with astronomers talking about gravitational collapse and experimentalists once again testing Einstein's predictions, it's as if the field of general relativity were waking up from a deep sleep. Theorists were renewing their interest as well.

The biggest hurdle for everyone at that time was describing a real star. Up to this point, all the work on "gravitationally collapsed objects" started with a ball of matter that was completely still. It was the only way that scientists, such as the groups at Princeton and in the Soviet Union, could solve the equations. But this was a very unrealistic simulation.

Stars spin. Every star in the sky rotates. So, there was still the possibility that a stellar collapse is avoided when rotation is taken into account. That was on the minds of many. They continued to believe that the infamous "singularity" was imaginary, simply an artifact of the way Einstein's equations were solved in that special case of a motionless star collapsing in a completely symmetric manner. Total collapse to "zero volume" still appeared too fantastical. But to prove that, relativists had to conquer their biggest unsolved problem: setting up the equations of general relativity in such a way that they could

Roy Kerr in 2013. (*Courtesy of the University of Canterbury*)

handle a rotating star. It was the field's holy grail. The solution eluded theorists for decades. That is, until Roy Kerr, a mathematical physicist and New Zealander, tackled it.

Right after World War II, Kerr completed his bachelor's and master's degrees at what is now the University of Canterbury in New Zealand, at a time when the library "was so bad that its modern physics books talked about ether theory," he recalled. Kerr became interested in general relativity after he moved to Cambridge University in England for his doctoral studies, where for his dissertation he considered how particles move under its rules, such as two stars closely orbiting one another.

By the start of the 1960s relativists were being reenergized with the introduction of new mathematical approaches from differential geometry to solving Einstein's field equations, which opened the door for more physics getting done. This new development created great excitement among relativists, shaking them out of their doldrums. Kerr got caught up in the fervor and began working on some of his own solutions, continuing as he settled into a job in 1962 at the University of Texas in Austin, where a new Center for Relativity was getting established.

It was tough going for Kerr, and after a few months of toil, a colleague in Austin showed him a paper about to be published that seemed to indicate that almost no solutions were possible for the problem he was working on. But skimming through the article, Kerr noticed a mistake in one of the equations, which indicated that wasn't true. "The next few weeks turned into a furious cocktail of adrenaline, trancelike bouts of distraction, and the smoke from seventy cigarettes per day," recounted physicist Fulvio Melia, Kerr's biographer. Kerr went on to reduce the problem to a set of "fourth order" differential equations, which agreed with the results made by another team of relativists, Ivor Robinson and Andrzej Trautman. Whereas those men were attempting to calculate the most general cases, Kerr at that point decided on a different strategy. He threw out any result that had no connection to the physical world. "I wanted to find a solution that could represent something we find in the universe," he says. He got rid of awkward terms by taking advantage of certain symmetries, a move that some thought inelegant. Most important, he chose a coordinate system that was axially symmetric. In other words, it had the potential to handle a rotation.

Kerr knew he was closing in when his solution matched Newton's law of gravity, in the case of an observer being far from the source of gravity. But from that perspective, the rotation was not obvious. The next day in his office, with his boss Alfred Schild sitting nearby in an old armchair in anticipation, the young mathematician sat at his desk with pencil and paper to verify that the object he had placed in his virtual space-time did indeed have angular momentum. After half an hour, chain-smoking cigarettes as he calculated, Kerr turned to his companion and uttered, "Alfred, it's spinning." More than that, the rotating object was dragging space-time around with it, like the cake batter that circulates in the bowl around a whirling beater.

Known as "frame dragging," this was a relativistic effect that two Austrian physicists, Josef Lense and Hans Thirring, first predicted using approximation methods in 1918. Kerr at last had the full solution. Schild was overjoyed by the result. "Cutting through the billowing smoke from his pipe," reported Melia, "[Schild] rushed to the desk and looked over Kerr's shoulder at the scratchings on the table." He immediately recognized that Kerr had at last found the way to refashion Einstein's equations to handle rotation. "I do not remember how we celebrated," recalled Kerr years later, "but celebrate we did!"

In the parlance of general relativists, Kerr had devised a new "metric" to describe the space-time around a spinning object with mass. He had conquered general relativity's Mount Everest of problems. The magnitude of this accomplishment was colossal—so much so that Kerr was quickly offered a tenured professorship at the university. It was quality—not quantity—that led to this outcome. His final paper, published in 1963 within a month of his submission to *Physical Review Letters,* was a mere one-and-a-half pages long. Schild was so excited by Kerr's accomplishment that he wanted the university to bathe its campus tower in a celebratory orange glow, just the way the college always did when its football team won a game. (It didn't happen.)

This all occurred just around the time that the Texas Symposium was being planned. Hearing that the symposium's organizers were scheduling someone else to discuss his solution at one of the sessions, Kerr made sure he'd be the one at the podium (though he may have regretted that decision). "It went over like a lead balloon," Kerr now recalls. In the few months between the publication of his paper and the Texas Symposium, Kerr had adapted his approach to handle an object collapsed to a Schwarzschild sphere. Given that quasars were the symposium's main focus, he wanted to show how his solution might "explain the large energies emitted by quasi-stellar sources in terms of the gravitational collapse of large masses." Rotation made a difference, he told the audience.

But the astronomers at the conference did not appreciate at all what the relativist had accomplished. They hardly listened while Kerr gave his ten-minute talk. Many slipped out of the auditorium to take a break; others snoozed in their chairs; a few ignored the speaker entirely and talked among themselves. The astronomers didn't think space-time metrics or Schwarzschild surfaces had anything to do with quasars.

The relativists in the room, however, were riveted. At the end of the talk Achilles Papapetrou, a noted Greek relativist, got up and declared that Kerr had arrived at the rotation solution that he and others had been trying to find for some three decades. He shook his fist and scolded the audience for not listening. In the greatest of ironies, the astronomers in the room effectively yawned at the news. Kerr that December day was handing astronomers on a silver platter what came to be understood as the first model of a spinning black hole. If clever astrophysicists at the meeting had listened and made the leap, they would have found a bit sooner another possible source for powering those quasars they were so excited about, the specific raison d'être of the meeting—tapping into a black hole's rotational energy.

A black hole's spin is one key to its power. Think of an ice skater, arms spread out, who then brings them in to spin faster and faster. It's the simple result of conservation of angular momentum: as the width of a rotating object decreases, its spin increases. A big rotating star that suddenly collapses to a small black hole takes this to the extreme; the black hole ends up spinning at humongous velocities. So fast, Kerr recognized, that the black hole develops two surfaces. The inner boundary is the standard event horizon, where any matter or light that crosses can never leave. But there's an outer boundary as well, spherical in shape but somewhat flattened so that it touches the black hole at its poles. Any light or matter entering the region between the two boundaries is whirled around at high speed and, if positioned in just the right way, has a chance to escape. The way out is along magnetic field lines, which direct the matter straight out of the black hole's north and south poles.

To be fair, Kerr didn't offer these details at the Texas Symposium. He didn't think of the region between the two boundaries in this way. He didn't even have his two surfaces defined correctly, due to his rush to get a result before the meeting began. But it was a start. And by 1969 Roger Penrose fully demonstrated how this special place between a black hole's inner and outer boundaries, which came to be called the ergosphere, can act as an energy amplifier. "Erg" is derived from the Greek word for work or energy, and that's exactly what the ergosphere provides. Penrose showed how any matter and light that enters into this special realm and then escapes actually *gains* energy—a lot of energy—from the black hole's rapid rotation. The loser here is the black hole, whose spin gets reduced a smidgen in the ergosphere process.

There was another outcome from Kerr's solution. For the staunchest opponents to gravitational collapse, rotation had always stood out as a last hope, a star's potential savior from total oblivion. But that was proven not to be true. Though rotation added some exciting new properties to a black hole, it didn't prevent the black hole from forming at all. Moreover, Kerr's solution was later proven by others, including Stephen Hawking, Brandon Carter, and David Robinson, to be the only type of black hole possible. Chandrasekhar called that discovery "the most shattering experience" of his scientific

A rotating black hole's two surfaces: the event horizon, from which nothing can escape once it enters, and the ergosphere, an outer region where it is possible to extract energy from the hole. (*Messer Woland, courtesy of Wikimedia Commons*)

life, the realization that Kerr's solution "provides the absolutely exact representation of untold numbers of massive black holes that populate the universe . . . that a discovery motivated by a search after the beautiful in mathematics should find its exact replica in Nature."

By the latter half of the 1960s, science fiction writers were fully alert to this new celestial kid on the block. In an episode called "Tomorrow Is Yesterday," which initially aired on 26 January 1967, during the first season of *Star Trek*, the starship *Enterprise* encounters an invisible "black star," as the narration put it, whose immense gravitational attraction dragged the spaceship dangerously close. Gravitationally collapsed objects were also routinely referred to as "dark stars," along with frozen stars and "collapsars." The expression we all know and love—black hole—was not made official until the end of 1967.

The term *black hole*, of course, has had a dark and notorious reputation. In June 1756, on the banks of the Hooghly River in Calcutta, India, at the British garrison of Fort William, 144 British men and 2 women were taken prisoner by the troops of the nawab of Bengal, Siraj ud-Daulah. According to one historian, Siraj's men incarcerated at least 64 of the hostages for a night in a tiny, cramped cell known as the "black hole." Few more than 20 reportedly survived the hot, suffocating night. Ever since that horrific event, the words *black hole* have referred to a place of confinement, a locked cell, where it was anticipated that once you went in, you never came out.

Wheeler repeatedly told the tale that he first used the term *black hole* in the fall of 1967 at a conference quickly set up at the NASA Goddard Institute for Space Studies in New York City after radio pulsars were discovered. Were the mysterious beeps coming from red giant stars, white dwarfs, neutron stars? According to Wheeler, he told

the assembled astronomers they might be his "gravitationally collapsed objects." "Well, after I used that phrase four or five times,
somebody in the audience said, 'Why don't you call it a black hole.'
So I adopted that," said Wheeler.

But, while pulsars were discovered in 1967, their existence remained a well-kept secret until 1968, the announcement held off until
that February, when the discovery paper was finally published by *Nature*. The pulsar conference at the Goddard Institute did not take
place until May. Perhaps Wheeler misremembered the conference
taking place in 1967. There was a meeting on supernovae at Goddard
in November 1967, but Wheeler's name is not found in the conference's proceedings. What is indisputable is that Wheeler used the
phrase during an after-dinner talk at the annual meeting of the American Association for the Advancement of Science (AAAS) in New
York City on 29 December 1967. It then made it into print when an
article based on that talk, titled "Our Universe: The Known and the
Unknown," was published in *American Scientist* in 1968. It was from
this publication that Wheeler is traditionally credited for the origin of
the term *black hole*.

Yet there's firm evidence that the term actually arose much earlier.
For one, it was casually bandied about at the 1963 Texas Symposium,
four years before Wheeler adopted it. The science editor for *Life* magazine at the time, Albert Rosenfeld, used the term *black hole* in an
article on the newly discovered quasars. Reporting on the Texas conference, he noted how Fred Hoyle and William Fowler suggested that
the gravitational collapse of a star might explain the quasar's energy.
"Gravitational collapse would result in an invisible 'black hole' in the
universe," wrote Rosenfeld. Rosenfeld today is sure he didn't invent
the term but overheard it at the meeting, although he could not recall
the source. Could Hoyle have used the term in his discussion? More

than a decade earlier, the British astrophysicist had wryly dubbed the explosive theory of the universe's origin as the "Big Bang." Was he using his talent for evocative astrophysical nicknames once again? Or were young graduate students and post-docs playfully using the term in the hallways of the conference?

The phrase was mentioned again a week later at an AAAS meeting held in Cleveland. Ann Ewing of *Science News Letter* reported that astronomers and physicists at the conference were suggesting that "space may be peppered with 'black holes.'" The person who used the term there was Goddard Institute physicist Hong-Yee Chiu, who had organized the session that Ewing covered and had also attended the Texas Symposium. Chiu had originated the term *quasar;* was he introducing another fun term to the public? No, answers Chiu; he borrowed it from the man who may have coined the phrase from the start.

From 1959 until 1961 Chiu was a member of the Institute for Advanced Study in Princeton, and during that time Princeton physicist Robert Dicke, both an experimentalist and theorist on gravitation, spoke at a colloquium and mentioned how general relativity predicted the complete collapse of certain stars, creating an environment where gravity was so strong that no light or matter could escape. "To the astonished audience, he jokingly added it was like the 'Black Hole of Calcutta,'" recalls Chiu. A couple of years later, when Chiu started working at the Goddard Institute, he heard Dicke casually use the phrase again there during a series of visiting lectures. In this way, Dicke may have released the term into the scientific atmosphere. It was one of Dicke's favorite expressions, for he often used it with his family in an entirely different context. His sons recall their father exclaiming, "Black Hole of Calcutta!" whenever a household item appeared to have been swallowed up and gone missing.

If Wheeler was unaware of these earlier uses, though, could he have been influenced by a poem titled "Music of the Spheres" written by A. M. Sullivan, which focused on the eighteenth-century astronomer William Herschel? The poem was published in the *New York Times* on 26 August 1967, just a few months before Wheeler's talk at the AAAS meeting in the city.

When the long eye of Herschel
Burrowed the heavens
Near the belt of Orion
He trembled in awe
At the black hole of Chaos.

Whoever inspired the phrase, Wheeler still deserves much of the credit for its placement into the scientific lexicon. Given Wheeler's status in the field, his decision to adopt the moniker bestowed a gravitas upon it, giving the science community permission to embrace the term without embarrassment. "He simply started to use the name as though no other name had ever existed, as though everyone had already agreed that this was the right name," said his former student Kip Thorne.

Wheeler's strategy worked splendidly. Within a year of his 1967 New York talk, the idiom was gradually being used in both newspapers and the scientific literature—although for a while it was first written down as the "black hole," an expression so exotic it needed to be held at a distance within quotation marks.

Some, like Richard Feynman, thought the term was obscene. "He accused me of being naughty," said Wheeler. But Wheeler was attracted to its link to other physics terms, such as blackbody, an ideal body that absorbs all the radiation that falls on it and is also the perfect emitter. A black hole does the former but not the latter. It emits nothing . . . zip. . . . nada. We look in and see only a dark emptiness. "Thus *black hole* seems the ideal name," concluded Wheeler. Moreover, it fit the very physics of the situation. The singularity, with its infinite density, was literally digging a hole—a bottomless pit—into the flexible fabric of space-time. And like some cosmic karma, the name also pays homage to the very man who started it all, Karl Schwarzschild. Schwarz means "black" in German.

"The advent of the term *black hole* in 1967 was terminologically trivial but psychologically powerful," said Wheeler. "After the name was introduced, more and more astronomers and astrophysicists came to appreciate that black holes might not be a figment of the imagination but astronomical objects worth spending time and money to seek." The black hole had finally made it into the big time. Its intriguing name provided the object with a beguiling personality it had lacked before.

Even Chandrasekhar returned to the subject, figuring it was now safe to get back into the game without ridicule. He had been away for nearly forty years, starting with his infamous kerfuffle with Arthur Eddington. And by the mid-1970s black holes were no longer the static objects that Chandra first encountered. Now they were active, twirling cosmic entities. After gulping down a large helping of matter, a black hole's event horizon can shake, rattle, and roll. Within eight years of his coming back to the topic, Chandra wrote one of the definitive books on the subject, *The Mathematical Theory of Black Holes*, which neatly packaged all the various techniques required to study a black hole's behavior. It remains a classic in physics departments to this day.

10

Medieval Torture Rack

Black hole studies prospered under the tutelage of Wheeler, Zel'dovich, Thorne, and others. It was at this point that all of a black hole's strange properties came to be identified and examined more deeply by theorists. Astrophysicists were catching "black-hole fever," and the infection spread rapidly. "There is a Chinese proverb. Ten years ago the river flows east, and ten years later the river flows west. Things change unexpectedly. Suddenly black holes became the hottest subject in town," says Hong-Yee Chiu today, who was once labeled a "crackpot" for being a supporter of neutron stars and black holes.

Fortune magazine, a purveyor of both business and research trends, noticed in 1969 "a notable migration of . . . scientists and graduate students into the fields of astronomy, astrophysics, cosmology, and relativity research." And general relativity was growing faster than almost any other area. By this point centers specializing in relativistic physics had been set up in Moscow, Paris, Syracuse University, the University of Maryland, North Carolina, Princeton, Berkeley, Caltech, and Cambridge University. The best students in physics were flocking to them. "Particle physics was in a mess at this time," recalls MIT physicist Alan Lightman, who under Thorne's guidance received his PhD in black-hole physics during this era. "There were dozens of different theories of the strong force, hundreds of new kinds of elementary particles, and no clarity at all. General relativity was more attractive because it was not yet glutted with practitioners. And, with the discovery of neutron stars in 1967, people were beginning to believe in the validity of highly compact stars, including black holes."

It was at this moment that magazines and newspapers began regularly publishing science stories filled with the frightening (yet strangely entertaining) consequences of being near a stellar-sized black hole. Black holes were proclaimed as "The Darkest Riddle of the Universe," "The Dazzling Death Spasm of a Star," "The Blob That Ate Physics," "Vacuum Cleaner of the Cosmos," and even the "Bermuda Triangles of Space." "Of all the concepts conjured up by physicists," wrote *New York Times* science editor Walter Sullivan in 1971, "none is more bizarre than that of the 'black holes' in outer space."

They became a cultural phenomenon as soon as they arrived on the public's radar, joked about on late-night television and in the press. A fake advertisement in the science fiction magazine *Analog* hawked "black-hole disposal units" in seven decorator colors that would suck up unlimited waste. T-shirts proclaimed that "black holes are out of sight."

Theorists, meanwhile, found humor in the physics itself. They jokingly talked about how you would get "noodlized" as you passed, feet first, through an event horizon. From the classic general relativistic perspective (more on another scenario—the quantum view—in chapter 12), once the horizon is crossed, you can't turn back. The only thing that lies ahead for you is the center of the black hole, whose gravitational pull increases as you plummet. In fact, the rise is so swift and sharp that the pull on your feet would be far greater than the pull on your head, so you'd get stretched out—just like a noodle. At the same time, you'd be crushed from side to side. It's the same effect—the tidal force—in which the Moon pulls on the Earth's oceans to generate the tides. But in the case of a black hole the tidal forces are gargantuan. In the blink of an eye, less than a millisecond for a stellar-size hole, your body would be broken up into cells, the cells into atoms, the atoms into elementary particles, the particles into quarks, and the quarks into entities yet to be figured out. Whatever the ultimate debris, all is swept into the infinitely dense singularity at the black hole's center and crushed to oblivion. Wheeler liked to describe this final entity, situated deep in the black-hole well, as "mass without matter."

Depending on the mass of the black hole, the timing of this scheme can change somewhat. Like a glutton's expanding waistline, the event horizon stretches outward, wider and wider, as a black hole devours more mass. If the total mass of the black hole is sufficiently large, you would not even realize when you've passed the point of no return. A black hole's event horizon is not a solid surface after all, but more like an invisible county or city line. Astronauts approaching the event horizon of a supermassive black hole, the kind that lurks in the hearts of most galaxies and contains the mass of millions or even billions of suns, would experience nothing but emptiness as they cross over. But eventually the tidal stretching will begin, as if in slow motion: head and feet pulled apart and the chest squeezed, as if he or she were on a medieval torture rack. For a black hole containing the mass of five billion suns, the fall from event horizon to final crunch would take the astronaut about twenty-one hours.

In lectures, Wheeler often liked to compare the distinction between the horizon and the crushing point to falling off a cliff onto some rocks below: "On first approach [the cliff] had sloped down safely, inviting the explorer to come closer. Unperceived in one's eagerness to peer over the edge, the slope of the slippery grass increased.

Then the shoes began to slide forward. Suddenly it became clear that disaster was inescapable, though it had not yet struck. That treacherous and unmarked point of no return symbolizes the equally treacherous and equally unmarked horizon of the black hole—as the rocks at the bottom symbolize the point of obliteration."

And why can't you escape? It's because the event horizon marks the point at which an object would have to accelerate to the speed of light, 186,282 miles per second, to break free—far faster than the mere 7 miles (11 kilometers) per second we need to leave Earth. Once you pass through the event horizon, there is no way out. You'd have to turn around and flee at a speed faster than the speed of light—an impossibility, according to Einstein. It would take an infinite amount of energy to do so. As a consequence, the black hole holds on to you with an iron grip.

When Einstein declared that space and time are relative, nowhere is this more apparent than in the realm of a black hole. Time slows down within a strong gravitational field, a natural outcome of general relativity that's been proven many times. You can think of each beat of a clock needing more time to climb out of its deep gravitational well. Indeed, as noted earlier, the clocks aboard the Global Positioning Satellites orbiting above us, whose signals help direct our driving and hiking, run just a tad faster than clocks down here on Earth, where the clench of gravity is more forceful. Black holes, the mightiest sinkholes in the universe, take this effect to the extreme.

Imagine you were miraculously able to sit on the surface of a collapsing star, just before it shrinks within its event horizon to become a black hole. Looking down at your watch, time is progressing normally. You see a second go by. Yet looking back out at the universe at large, billions of years are elapsing. The future history of the cosmos is rushing past at near-light speed. Anyone watching you from afar,

An illustration of a stellar-sized black hole, as seen from a distance of about four hundred miles (644 kilometers), with the stars of the Milky Way in the background. The starlight is distorted and stretched as it gets bent by the black hole's intense gravitational field before reaching our eyes. (*Ute Kraus, Universität Hildesheim, courtesy of Wikimedia Commons*)

however, sees something vastly different. Removed from the black hole's immense gravitational field, the distant observer thinks you are taking an eternity to cross the event horizon. To you, of course, that's not the case at all. In your time frame, you would instantly die. But to your observer, you appear to be motionless at the brink of the event horizon—forever young and forever saved from total destruction. From the perspective of your far-off observer, time near the black hole has practically ground to a halt. No wonder Russian theorists at first preferred to call the black hole a frozen star. In reality, though, this star would still look black to a distant onlooker, as the last light waves escaping from it are stretched silly to infinite lengths, rendering them invisible to our eyes.

The notion of the black hole as a frozen star affected astrophysicists' beliefs for quite a while. It made them assume that a black hole would have no influence whatsoever on our present-day universe. From our reference frame in time, the star was essentially a petrified object, so why should it affect us? "As long as this viewpoint prevailed," noted Richard Price and Kip Thorne in a book on black holes, "physicists failed to realize that black holes can be dynamical, evolving, energy-storing and energy-releasing objects."

Astronomers were just starting to learn that: first with quasars, and later in discovering stellar black holes within our own home galaxy.

11

Whereas Stephen Hawking Has Such a Large Investment in General Relativity and Black Holes and Desires an Insurance Policy

Setting up an astronomical observation to look for a black hole within our galaxy took some time. Many had to be convinced they were not only worth seeking out but even possible to hunt down. For Oppenheimer back in the 1930s, black holes were strictly a theoretical problem. He wasn't going to bother to look. Why should he? He'd proven that his gravitationally collapsed object would vanish from sight. Astronomers at the time weren't interested for other reasons. They thought all this talk about dying stars collapsing was foolish. Stars simply didn't behave like that.

Wheeler, in contrast, did come to imagine neutron stars and black holes as real denizens of the universe. But even he was short-sighted at first about the possibilities of finding them. In 1964, he published an article for the book *Gravitation and Relativity* on the neutron star, what he called the "superdense star," in which he said, "There is about as little hope of seeing such a faint object as there is of seeing a planet belonging to another star." (Of course, *both* are now observed regularly by astronomers.) But this was 1964, when such technological abilities were hardly even fantasies. And no one had yet conceived that a neutron star could emit intensive radiation from its poles as it wildly spins—radiation detected by us across the electromagnetic spectrum, from radio to X-rays.

Zel'dovich and his team, however, were already thinking about this problem deeply. How do you make the invisible, well, visible? How could you possibly detect a pitch-black object journeying through the darkness of space? At first, they borrowed an idea that John Michell had mentioned back in the eighteenth century: look for a luminous star that is wiggling back and forth because a dark companion is tugging on it as it orbits. If that companion emits no light and gravitational measurements show it weighs a few solar masses or more, then it could very well be a black hole. Zel'dovich drafted an astronomy graduate student, Oktay Guseynov, to comb the binary-star catalogs to come up with candidates. By 1966 they found five. (In their report to the *Astrophysical Journal,* they inserted a little jab against those astronomers who continued to insist that stars lose enough mass to avoid collapse. Yes, the Soviet researchers conceded, stars can shed matter but "not because they 'wish' to be white dwarfs or 'fear' to be collapsed.") Working with astronomer Virginia Trimble, Kip Thorne later came up with eight more black-hole candidates. But in the end, none of these turned out to be viable contenders. There are many reasons that a companion could be very faint and still not be a black hole. What was needed was a new tool altogether to conduct a search.

Fortunately, Zel'dovich and his colleague Igor Novikov realized within a few years that there was another way to unmask a black hole, a variation of the accretion process they had already mentioned for quasars. Imagine a black hole in orbit around a brightly glowing star, whose surface is releasing streams of gas in a stellar wind. The black hole might even be tugging away its companion's outer atmosphere. Eventually, some of that gas will reach the black hole and be captured by its powerful gravitational field. As this gas plummets toward the hole, with its atoms churning and colliding, it is heated to millions of degrees. And in the process, it emits copious radiation, not as visible rays but as *X-rays*. Though black holes are invisible against the dark backdrop of space, they would give themselves away by how they affect their surroundings. A black hole would announce itself by the glow of fiercely energetic X-rays enveloping it (just as with quasars). "[This] proposed method of searching for 'black holes' in binaries," Zel'dovich and Novikov later wrote, "reminds one of the well-known case of searching for a lost key under a lamppost: the key is sought where it is easier to find." Seeking out bright X-ray sources is less painful than searching through piles of stellar catalogs for wiggling stars. Fortunately, around the same time the two Soviet researchers arrived at this realization, X-ray astronomy was fast maturing as a new means of studying the universe.

X-ray astronomy was almost doomed from the start. Soon after World War II, researchers from the US Naval Research Laboratory used surplus German V-2 rockets to loft instruments high above our atmosphere to capture X-rays emanating from the Sun. Such rays are impossible to detect on the ground. Though they can easily penetrate matter over short spans, they are completely absorbed in our broad atmosphere, being very short electromagnetic waves that span the width of an atom. Although the solar X-rays these pioneers detected were very intense from a layperson's point of view, the output was relatively meager by cosmic standards. So, making estimates based on that 1948 discovery, theorists figured that X-ray emissions from faraway stars would be minuscule by the time they reached Earth, a billion times fainter than the Sun's X-ray output. Since such a signal was impossible to detect with 1950s technology, it didn't seem worth pursuing.

But American desires to better monitor Soviet nuclear bomb tests from space in the early 1960s accelerated efforts in improving X-ray detectors. Such bombs release appreciable X-rays. A twenty-eight-year-old named Riccardo Giacconi, who arrived in the United States as a Fulbright scholar from Italy and stayed to serve as a physicist with the private research firm American Science and Engineering, headed up the effort. His team's newly built instruments worked nearly perfectly when measuring a series of American bomb tests. It wasn't long before they turned those same detectors to the heavens.

One particular rocket flight dramatically set the stage, initiating the birth of X-ray astronomy. Under the light of a full Moon, the launch took place one minute before midnight on 18 June 1962. Giacconi, along with his colleagues Herbert Gursky, Frank Paolini, and Bruno Rossi, had mounted their payload, three large Geiger counters, onto a small Aerobee rocket and launched it from the White Sands Missile Range in southern New Mexico. After reaching an altitude of 140 miles (225 kilometers), the rocket plunged back to Earth. Two of the detectors, sweeping the sky as the rocket spun on its long axis twice a second, recorded 350 seconds of useful data. They were six of the most fruitful minutes in astronomical history.

This was when the United States was gearing up for its longtime goal to land a man on the Moon by the end of the decade. Giacconi and his colleagues were trying to detect X-rays from the Moon. They figured the radiation would be generated as the energetic solar wind struck the lunar surface and were hoping that the X-ray spectrum would help them determine the Moon's composition. But Giacconi and Rossi had also long suspected that extrasolar X-rays would be emanating from such celestial objects as supernova remnants. We were "trying to get support . . . frankly wherever we could," said Giacconi. Unable to garner a NASA grant to look for X-rays from deep space, the two cunningly used the US Air Force–funded Moon test as an opportunity to take a look around the celestial sky while the rocket was in space.

No lunar X-rays were detected during the brief flight, but the rocket team hardly despaired. They found something far more enticing, the kind of novel signal they had been hoping for all along. The rocketborne detectors noticed a huge flux of X-ray radiation arriving from a region of the sky in the direction of the constellation Scorpius. Hence, the name it eventually received: Sco X-1, as it was the first X-ray source in that sector of the sky. Sco X-1, located some nine thousand light-years from Earth, blazed with an X-ray intensity beyond anyone's imagination, tremendously more powerful than our Sun's meager output. It was millions of times stronger than the X-ray radiation emitted by normal stellar sources. This was such a leap in strength that at first the team worried their instrument had been damaged, producing a false signal. Others in the scientific community were wary as well and demanded confirmation.

Proof arrived when further rocket flights found additional sources similar to Sco X-1. And starting in 1970, as soon as astronomers placed X-ray-detecting satellites into orbit, many more were found. With these advanced tools, astronomers learned that many of these energetic sources are neutron stars in binary systems. The X-rays are released as matter from the normal, visible star is drawn away and funneled onto the surface of its superdense neutron star companion. Many of these neutron stars appear to regularly pulse in X-rays, as the neutron star rapidly spins. The X-ray "hotspots," formed at the poles where the matter is magnetically drawn, go in and out of view much like the rotating lamp of a lighthouse. This was an important turning point. With the mounting evidence that neutron stars populated the galaxy—both as radio pulsars and as X-ray-emitting sources—it became easier for astronomers to take the plunge and accept the possibility that black holes existed as well. "The discovery of pulsars," noted Kip Thorne at the time, "opened the floodgates." Astronomers were at last willing, he said, "to take seriously some of the wildest meanderings of theoreticians' minds, including speculations about the role of black holes in the Universe."

This played out when one of the X-ray sources appeared to be in a class by itself. Fully revealed in 1964 during one of the continuing series of rocket flights, it was labeled Cygnus X-1, signifying its location in that constellation (also known as the Northern Cross). By then, other research groups were joining in the hunt with their own rocket launches, and each proceeded to measure a different X-ray brightness for this source, which puzzled everyone. A long look by the first X-ray satellite, *Uhuru*, in 1971 finally disclosed that this luminous patch in Cygnus was atypical. Rather than emitting regular pulses of X-rays like the other sources, it underwent sporadic variations. There was no discernible pattern to its flare-ups. Sometimes its signal flickered over periods as short as millionths of a second, a sign that whatever was giving off those X-rays had to be fairly compact. If the source was as big as a normal star, the radiation pulse would have lasted much longer.

At the March meeting that year of the American Astronomical Society, then being held in Baton Rouge, Louisiana, Giacconi boldly suggested Cygnus X-1 might be a black hole. With so many neutron stars being found with more confidence, the thought of black holes existing became easier to contemplate. The day after Giacconi's announcement, a headline in the *New York Times,* splashed across the top of page 20, proclaimed, "An X-Ray Scanning Satellite May Have Discovered a 'Black Hole' in Space." Notice that quotation marks were bracketing the term as late as 1971. The object still seemed too strange to be true.

Follow-up work by both radio and optical astronomers at last pinpointed the source. Observers determined that Cygnus X-1's powerful X-rays were coming from a double-star system, in which a giant blue star (with the rather mundane tag HDE 226868, for its number in the Henry Draper extended star catalog) is coupled with a dark, invisible companion. The blue supergiant closely orbits its unseen partner once every 5.6 days, a tempo that allowed astronomers to apply Newton's laws and determine that the unseen companion must have a mass beyond that of our Sun. By late 1972 orbital measurements suggested that the mass was at least ten times greater than our Sun's—too massive for a neutron star and so a prime candidate to be a black hole. (Current estimates put its weight at around fifteen Suns.) Invisibility plus sizable mass—coupled with the rapidity of its X-ray fluctuations suggesting a tiny size—added up to *black hole* with far more surety. It became astronomy's first prime suspect.

If you could somehow hover above Cygnus X-1, situated about six thousand light-years away, you would witness a gaseous whirlpool of enormous dimensions. Observations suggest that the black hole is pulling matter off of its generously endowed companion and forming a disk of gas around itself. This disk is flattened due to both centrifugal and gravitational forces. Like satellites orbiting the Earth, this material does not fall straight into the hole. Instead, it orbits the space-time drain in tighter and tighter spirals. Wheeler once compared it to traffic converging on a sports stadium from all directions and becoming more and more closely packed as the cars approach their destination.

And this has definite consequences. As the gas gets squeezed further and further, its temperature rises tremendously. Heated to tens of millions of degrees, the hot gases start to emit copious amounts of X-ray energy. This is the radiation that X-ray telescopes spy before the matter is drawn into the gravitational abyss and lost to our view. It

An illustration of the black hole Cygnus X-1 stealing gas from the atmosphere of its companion star. This gas first orbits the space-time drain, emitting in the process large amounts of energy, before it is eventually swallowed by the black hole. (*NASA/CXC/M. Weiss*)

may take weeks or even months for any one blob of gas to travel the few million miles from the outer edges of the disk to the point of no return. But in its last moments, the gas swirls around the hole thousands of times each second, possibly causing those rapid X-ray fluctuations.

Of course, this scenario took time to formulate and prove. The first claim that Cygnus X-1 might be a black hole was fraught with peril. "Show me the evidence" was the cry on nearly every astronomer's lips. Much of the proof was circumstantial, more a game of connect-the-dots than irrefutable confirmation.

This made the prospect both exciting and controversial, so much so that Stephen Hawking and Kip Thorne made an infamous bet at Caltech in December 1974 on whether Cygnus X-1 was truly a black hole. Hawking bet against, Thorne for. The agreement, handwritten on one page of stationery and with racy US and British magazines at stake, declares:

> Whereas Stephen Hawking has such a large investment in General Relativity and Black Holes and desires an insurance policy, and whereas Kip Thorne likes to live dangerously without an insurance policy,
>
> Therefore be it resolved that Stephen Hawking bets 1 year's subscription to "Penthouse" as against Kip Thorne's wager of a 4-year subscription to "Private Eye", that Cygnus X 1 does not contain a black hole of mass above the Chandrasekhar limit.

Given his more generous bet, Thorne appeared to be four times more confident.

Verification was slow, but advances in X-ray astronomy helped. The *Uhuru* satellite was succeeded by astronomy's first true X-ray telescope in 1978. Instead of radiation counters that simply registered signal strengths, the spaceborne *Einstein* observatory housed a set of nested mirrors that actually focused the X-rays, allowing the radiation to be recorded as images as clear as any ground-based optical telescope. By 1990, according to Thorne, Cygnus X-1 was looking more and more like a black hole, with 95 percent confidence. That was a high-enough threshold for Hawking to cry uncle. "Late one night in June 1990, while I was in Moscow working on research with Soviet colleagues," recounted Thorne, "Stephen and an entourage of family, nurses, and friends broke into my office at Caltech, found the framed bet, and wrote a concessionary note on it with validation by Stephen's thumbprint." To the dismay of his wife, Carolee Winstein, Thorne won the *Penthouse* subscription.

For some, even stronger evidence for black holes had arrived from the far universe, as astronomers continued their examination of both quasars and radio galaxies with every spectral weapon in their arsenal: optical, X-ray, and especially radio.

In optical photographs, radio galaxies can appear quite boring. But radio telescopes, as mentioned earlier, revealed them to have a bewildering architecture. Astronomers saw that the visible galaxy is but a smudge caught between two sizable lobes of radio emission. Looking like a pair of water wings, these lobes stretch out for hundreds of thousands of light-years beyond the edges of the visible galaxy. By the early 1970s, a number of British theorists, including Martin Rees and Roger Blandford, concluded that some kind of plasma beams, monstrous ones at that, had to be responsible for pumping energy out into the lobes.

The desire to locate this river of plasma spurred countries to build ever bigger radio telescope networks, such as the Very Large Array (now known as the Jansky Very Large Array), twenty-seven radio dishes aligned in the shape of a Y over the plains of New Mexico. Together they can simulate a single radio telescope as large as the city of Dallas. With the increased power and resolution of such arrays, astronomers confirmed what the British theorists had suspected: the radio images displayed a sort of umbilical cord running from the nucleus of a radio galaxy out to its lobes: two thin beams of energetic, charged particles, each shooting out of the galactic core in opposite directions, at speeds of tens of thousands of miles per second.

A powerful jet of electrons and subatomic particles streaming from the supermassive black hole situated in the center of the giant elliptical galaxy M 87, situated some fifty million light-years from Earth. (*NASA and the Hubble Heritage Team at the Space Telescope Science Institute and AURA*)

Like the fierce stream of a fire hose, these cosmic jets can bore through the thin gases found in intergalactic space, until they ram into a denser region of gas, as if the stream had come up against a brick wall. The particles in the jets then fly off, filling up the gigantic lobe regions.

The next question was a natural: What could possibly keep these cosmic jets flowing? Theorists agreed the engine had to be pretty special. First of all, the power source had to be fairly stable, in order for the jets to maintain their orientation over millions of years. And radio "pictures" were getting so good that they were able to zoom into the very heart of the galaxy's core, showing a tiny spot whose brightness could fluctuate over days or weeks, which suggested that the engine was as small as our solar system. Moreover, this power source had to somehow eject its energy into two, oppositely directed beams.

There was only one power source that fulfilled all the design specifications: a spinning black hole formed from the collapse of millions, even up to billions, of suns. These suns may have first huddled together as an extraordinarily dense herd of stars, a type of cluster that could easily have developed early on in a crowded galactic center. Moreover, these first-generation stars, formed out of the pristine hydrogen and helium forged in the Big Bang and devoid of the heavier elements made later, were likely very big: so massive that they lived fast and died young as black holes. Driven by the inward force of gravity, these many holes could have ultimately coalesced into one giant black hole, which continued to grow over the eons as it "ate" any available stars or gas that got too close.

Or maybe a host of "baby" galaxies, building blocks in essence, coalesced into a bigger galaxy and during the turbulent chaos of this merger directed huge amounts of gas toward the center, which accumulated to incredibly high densities—so dense that the gas didn't turn into stars but directly collapsed into a massive black hole, serving as the seed for the supermassive black hole as it continued to grow and grow. The ultimate size of the supermassive black hole appears to depend on the mass of the galaxy's central bulge. Astronomers have found a direct correlation: the higher the bulge mass, the more massive the central black hole.

Whatever the giant hole's origins, theorists soon recognized that such an object was the most efficient energy generator for an active galaxy. When matter is thrown down a deep gravity well, particles are accelerated to velocities near the speed of light. Such gravity-driven motors can generate up to a hundred times more energy than nuclear-fired engines.

You might ask, "But doesn't a black hole completely eat up everything that enters it? How does any energy survive to get out?" The answer lies in picturing the environment surrounding the black hole. Throughout the 1960s, as noted earlier, a number of theorists, including Yakov Zel'dovich, Igor Novikov, Edwin Salpeter, and the British astrophysicist Donald Lynden-Bell, realized that as stars and gas are drawn in by the powerful gravitational pull of the black hole, it will form a doughnutlike ring around itself. Like water building up and swirling around a drain, the plummeting gas forms an "accretion disk" around the supermassive black hole, like the one described earlier circling Cygnus X-1. This disk rotates around, and in the same direction as, the spinning black hole. Enormous amounts of energy can then be released as this maelstrom of matter spirals inward toward the black abyss and is ripped apart by the gravitational tug-of-war. Excess gas that has not yet reached the hole's infamous point of no return might get magnetically deflected—funneled out like a cream filling from the top and bottom of the doughnut. That could be one source of the jets.

But there's another way as well. It's likely that an appreciable amount of the power is tapped from the spin energy of the galaxy's supermassive black hole. The black hole, in this case, is an electrical dynamo, but one of cosmic proportions. In this scenario magnetic lines of force, originating from the disk's gas, thread through the spinning hole's outer surface and whirl around with its rotation. Because of the hole's incredible spin, the magnetic field lines come out of the north and south poles of the hole coiled like streamers around a maypole. This forms two narrow yet powerful channels. Like a gigantic turbine in a galactic power plant, these spinning fields produce a huge electrical potential, which generates beams of particles that shoot out along each channel at near the speed of light (a model originated in 1977 by the British-born theorist Roger Blandford, now at Stanford University, and Roman Znajek). In this way energy is extracted from the rapid spin of the hole. It's the most efficient mechanism now known in the universe for converting matter into energy.

The spin also allows the black hole to act like a gyroscope, a device that has the ability to maintain a fixed orientation. That's how the cosmic jets can be steadfastly pointed in the same direction over long stretches of time, never varying. While this model has been tweaked and adjusted by many theorists over the decades, it had its roots in Roy Kerr's solution to Einstein's equations of general relativity, which was the first to demonstrate the behavior of a rotating object on the fabric of space-time.

Astronomers now see an evolutionary link between the quasars (those active supermassive holes) of yesteryear and the galaxies of today. As observers gaze farther and farther back in time, they count more and more quasars with very high luminosities. That's because the universe was spanking new and vibrant, constructing galaxies full of newly formed stars that were surrounded by lots and lots of gas. Under such conditions, the supermassive black hole building up within each young galaxy was able to gobble up its rations like a chowhound at an all-you-can-eat buffet.

But such a food supply is finite, and the black hole can haul in material only within a certain distance. As astronomer Richard Green once put it, "The gas gauge says 'empty,' and there's no gas station in sight." So, after some ten million to a hundred million years have gone by, a blip in cosmic history, the quasar's fireworks eventually taper off or simmer down to a less active state. It turns into a rather normal-looking galaxy. It was once thought that quasar activity was somewhat rare, but astronomers now believe that every sizable galaxy with a central bulge has a supermassive black hole at its center, an ancient quasar that could be triggered once again. A fairly common-place galaxy, for instance, can transform into a bright active galaxy or strong radio galaxy, loudly broadcasting its presence, after it collides with another galaxy, which causes new supplies of gas to feed the drowsing monster that had been sitting restfully in the galaxy's center.

The heart of our own Milky Way galaxy harbors a former quasar, a dormant supermassive black hole. A global team of radio astronomers is now readying a massive effort to image the "shadow" this hole casts against the brighter gas emissions surrounding it. This black hole, estimated to contain the mass of around four million Suns (on the small side, compared to galaxies with billion-solar-mass black holes), is now idling at low gear. Its engine does rev up occasionally, whenever it can grab some nearby fuel (say, a gas cloud falling into it), but that activity is peanuts compared to what may happen some four billion years from now. That's when the behemoth could fully re-awaken and roar loudly as our spiraling home slowly collides with our close neighbor, the Andromeda galaxy, whose central black hole is roughly ten times more massive. By the end of this fateful meeting, the two galaxies will combine to form a giant elliptical galaxy. Their holes will merge as well, putting on a breathtaking show of activity as the newly combined hole gulps down fresh sources of gas, released in the collision, and continues to grow to at least one hundred million solar masses.

12

Black Holes Ain't So Black

The portrayal of the black hole described in these pages so far is not complete. The behaviors depicted—strange as they may sound—have been rooted in a very classical mathematical scheme, general relativity. What hasn't been taken into account is quantum mechanics. What would a black hole look like from the perspective of an atom? And therein lies the rub: no one has successfully formulated a quantum theory of gravity. It's been done for all the other forces: electromagnetism, as well as the weak and strong nuclear forces. Gravity so far has been left out. It is the last, great task of theoretical physics—to fully merge general relativity with quantum theory.

There's a reason gravity's been the odd man out. Where all the other forces involve particles that follow the probabilistic rules of the quantum world—allowing these forces to be united into one grand mathematical scheme—general relativity's key parameter (at least in the way Einstein formulated it) is geometrical: curvatures in space-time. It's as if nature has set up two different sets of rules—one for gravity, and another for all the other forces. The tools that work so well in one realm are difficult to apply in the other. Gravity and quantum mechanics don't easily share the same mathematical vocabulary.

Despite these difficulties, several researchers in the 1950s and 1960s felt that the best way to energize the field as general relativity was waking back up from its decades-long slumber was to restart an effort that had begun in the 1930s—to bring quantum effects into general relativity. A number of notables, including Paul Dirac, Richard Feynman, and Bryce DeWitt, pioneered this effort, showing how gravity could be described in another way. Everything in the quantum mechanical universe—energy, motion, spin, and so forth—comes in indivisible bits. Forces fit naturally into this framework. Instead of viewing magnetism, say, as the result of invisible lines of force emanating from a magnet, the quantum world transforms the notion of force into an exchange of force particles—a subatomic tennis game. In electromagnetism this diminutive tennis ball is the photon, a particle that constantly bounces between charged particles, generating a force of either attraction or repulsion. By applying this same principle to gravity, the force of attraction between masses is conveyed by the continual transmission and absorption of "gravitons," particles that exist, for now, only hypothetically; they have not been detected.

But there's a big problem in recasting gravity in this way. Theories that treat forces as particles assume that every event in the subatomic world takes place on a fixed, unchanging background of space and time. Space-time is the stage upon which the actors, particles such as photons, flit to and fro. Space-time is not a participant. But in general relativity the distinction between stage and actor doesn't exist. According to Einstein, gravity is the very geometry of space-time. Thus the graviton becomes both actor and stage simultaneously. A graviton enters onto the stage of space-time, but by doing so ends up bending and warping the stage as if it were so much Jell-O. Theorists involved in solving this conundrum are far from arriving at a complete and unified solution.

But the black hole offered the means to gain a new perspective on this problem. It occurred when a few young up-and-comers in physics extended their analysis of a black hole's properties. Stephen Hawking was one of them. Diagnosed at the age of twenty-one with amyotrophic lateral sclerosis, also known as Lou Gehrig's disease, he had been expected to live no longer than two or three more years. He has now beat the odds by half a century. As an undergraduate at Oxford, Hawking was recognized as brilliant but underchallenged. He himself admits that it was his ensuing illness—the possibility of an early death—that put an end to his academic laziness.

Moving to Cambridge University for his doctoral studies, Hawking chose to specialize in cosmology at a time when it was more speculation than science, a risky choice. Yet, on obtaining his PhD in 1966, success was swift. First, he proved not only that the Big Bang didn't just *appear* to emerge from an infinitely dense point of mass-energy but that there was no other way. Then he discovered a vital link between gravity and quantum mechanics, two fields completely incompatible before then. As he reported in his best-seller *A Brief History of Time,* this particular insight was initiated as he was turning in for the night. "One evening in November [1970] . . . I started to think about black holes as I was getting into bed. My disability makes this rather a slow process, so I had plenty of time."

Continuing these mental ruminations, he eventually proved that a black hole's event horizon must always increase—never decrease—when matter falls into it. This might seem obvious, given that a black hole by definition allegedly never gives anything back, but until then it was mathematically fuzzy. As Hawking put it, "There was no precise definition of which points in space-time lay inside a black hole and which lay outside." Hawking defined it. He announced his discovery the following month at the Fifth Texas Symposium of Relativistic Astrophysics, held that year in Austin. The session devoted to black hole research was unexpectedly popular, attracting so many people that the organizers had to move it to a bigger auditorium.

The fact that a black hole's surface area must always increase looked a lot like the law of entropy in classical physics. Entropy is the measure of a system's disorderliness, how jumbled up it is. The higher the entropy, the greater the disarray. If things are left to themselves, entropy *always* increases. A rigid ice cube will melt and form a messy puddle, but without a refrigerator on hand to provide the energy, the untidy water can't reorder itself into ice on its own. It remains a patchy pool of water. Similarly, a black hole can only increase the size of its event horizon as it gulps down more matter. It can never decrease its girth. But Hawking and his colleagues took this resemblance between entropy and the size of a black hole—the fact that both properties are constrained to increase only—as merely a nice analogy. It didn't mean they were actually related in any way.

But one of John Wheeler's students, Jacob Bekenstein, boldly decided that the connection *was* real—that the area of an event horizon was indeed a direct measure of the black hole's entropy. On the face of it, this seemed like an outlandish proposition. From a classical perspective, a black hole is highly ordered. It gravitationally attracts everything within its reach and never gives its back. In fact, some wondered whether a black hole's entropy was *zero,* meaning it was the highest organized state possible. Wasn't all that matter getting compressed into an infinitesimally small point?

Because of this understanding, some warned Bekenstein that his research was headed in the wrong direction. But, as he later recalled, "I drew some comfort from Wheeler's opinion that 'black hole thermodynamics is crazy, perhaps crazy enough to work.'" It certainly attracted a lot of attention. There was a packed house when Bekenstein gave a seminar on his ongoing work at the University of Texas. "I suspect . . . ," he wrote Wheeler, "that the excellent attendance was due not so much to the particular topic, but rather to the great glamour of black hole physics in general. Yes, black holes are the hottest things in physics (and astronomy), not least thanks to your own early efforts in spreading the idea."

As Bekenstein continued to work out his calculations, the young student eventually came to see that the black hole would also have a temperature. But this is where Bekenstein drew the line. Even he tiptoed away from that comparison at first. It was universally accepted that a black hole holds on to everything it swallows. It emits nothing. So, it couldn't have a "temperature." Not a real temperature. That would imply that it was releasing radiation that could be measured by us as heat. "Such an identification can easily lead to all sorts of paradoxes, and is thus not useful," concluded Bekenstein in his 1973 published paper. So, all the top theorists declared that the temperature of a black hole was "unambiguously zero."

Stephen Hawking thought so, too. He was highly skeptical of Bekenstein's black-hole-has-entropy scheme and planned to publish a paper proving its conclusion was wrong. "I was motivated partly by irritation with Bekenstein," said Hawking in *A Brief History of Time*. Hawking felt that Bekenstein was misusing the earlier paper Hawking had written on the increasing area of an event horizon. "However," admitted Hawking, "it turned out in the end that he was basically correct."

Hawking was initially doubtful because anything with a large entropy should also be radiating. But a black hole, by its very definition, doesn't let anything out of its grip. Or does it? The more Hawking looked into this problem, the more he was intrigued, leading him to one of his greatest theoretical triumphs.

Hawking's outlook changed when he began looking at the black hole from a different perspective: from the viewpoint of an atom. His musings along this line were sparked during a visit to Moscow in the

Stephen Hawking (*American Institute of Physics Emilio Segrè Visual Archives, Physics Today Collection*)

fall of 1973, where he talked with Yakov Zel'dovich and the Russian's graduate student Alexei Starobinsky. These two men suggested that under special circumstances—that is, when a black hole rotates—it should convert that rotational energy into radiation, thus creating particles. This emission would continue until the spinning black hole wound down and stopped turning.

Devising his own mathematical attack on the problem, Hawking was surprised to discover that *all* black holes—spinning or not—would be radiating. As Hawking later put it, in one of the chapter titles of his popular book, "Black holes ain't so black."

Hawking announced his discovery in February 1974 at a symposium on quantum gravity, held at the Rutherford Laboratory near Oxford. His report was soon published in the journal *Nature* on March 1. Both his talk and paper were titled with the intriguing question, "Black Hole Explosions?" There was a reason that he mentioned explosions. In applying the laws of quantum mechanics to a black hole, Hawking found that black holes create and emit particles as if they were hot bodies. As a consequence, the black hole slowly decreases in mass and eventually disappears in a final blast! Such a finding turned black-hole physics upside down; a black hole, by definition, holds on to everything it swallows. It's supposed to emit nothing and never go away.

Hawking estimated it would take far longer than the age of the universe for a regular black hole, weighing a few stellar masses, to evaporate entirely. For a stellar-mass black hole (or larger), such a decay would take more than 10^{66} years. But what if extremely small holes were created in the turbulence of the Big Bang, each containing 10^{15} grams or so, about the mass of a small mountain? They could be popping off right now. Hawking estimated that in its final breath—its last tenth of a second of life—such a "tiny" object would release the energy of one million megaton hydrogen bombs.

Needless to say, this idea did not enthrall his fellow physicists. Relativist Werner Israel says that it "aroused strong opposition almost as soon as it was in print. . . . Skepticism was prolonged and virtually unanimous." When Hawking first announced his result at the February conference, it was greeted with total disbelief. At the end of the talk the chairman of the session, John Taylor from Kings College, London, claimed it was all nonsense. "Sorry, Stephen," he said, "but this is absolute rubbish."

But gradually, over the following two years, it came to be recognized that Hawking had made a startling breakthrough. "I was probably the most pleased," said Bekenstein, "for it provided the missing pieces of black-hole thermodynamics." The temperature of a black hole was not zero at all; it was the temperature of the radiation coming out of the hole, now known as "Hawking radiation."

Hawking arrived at his conclusion by asking how a black hole might affect its surroundings on the submicroscopic scale. He concluded that space-time gets so twisted near a black hole that it enables pairs of particles (a nuclear particle and its antimatter mate) to pop into existence just outside the black hole. You could think of it as energy being extracted from the black hole's intense gravitational field and then converted into matter.

But because we're talking about physics on the tiniest of levels, the exact boundary line of the event horizon is now quite vague. So at times, one of the newly created particles can disappear into the black hole, never to return, while the remaining one remains outside and flies off. As a result, the hole's *total* mass-energy is reduced a smidgen. This means the black hole is actually evaporating! Ever so slowly, particle by particle, the black hole is losing mass.

For stellar-size holes, this bizarre quantum-mechanical process is just about meaningless. As noted earlier, it would take trillions upon trillions of years for a regular black hole to shrink away to nothingness. Its temperature, gauged by the radiation coming out of it, would be less than a millionth of a degree above absolute zero. But Hawking suggested that the early universe, in the first turbulent moments of the Big Bang, might have manufactured a multitude of tiny black holes: mini–black holes. Like a ball rolling down a hill, the evaporation of such a mini–black hole would accelerate as time progresses. The more mass this tiny primordial object loses, the easier it is for the particles to escape. The hole ends up fizzling away faster and faster, until it reaches a cataclysmic end.

If the Big Bang did forge some mini–black holes, the smallest would have vanished before their dying light could catch our attention, but objects containing that mass of a mountain, yet compressed to the size of a proton, would be shedding the last of their mass at this very moment in a short and spectacular burst of gamma rays. No signals from such tiny holes have been detected with absolute certainty as yet, but astronomers continue to be on the lookout for that distinctive *pop*.

There is more to this story. Hawking's revelation initiated an entirely new examination of the black hole and opened up questions about the known laws of physics. Bizarre as its behavior may seem, the black hole originated within classical equations of physics. Einstein's general theory of relativity used the math of the nineteenth century, with space-time as his fundamental quantity. From that viewpoint, the black hole is a smooth and unbroken pit in the fabric of space-time. The event horizon is a point of no return, but there is no visible change in space-time at the transition. But Hawking showed that the black hole had an entirely different personality when examined on the submicroscopic scale. The event horizon is no longer smooth but more fuzzy and indistinct, as particles boil out of the vacuum—even violently as the black hole ages. It made physicists take a real, hard look at the black hole. Which was the real black hole? The version that arises out of Einstein's theory or the quantum mechanical one? How could these two completely different views be reconciled?

For a while, some wondered whether the rules of quantum mechanics are altered within a black hole: that quantum effects, inside the event horizon, are somehow different from those we measure on the outside. But the physicists who are examining this cutting edge of black-hole science are coming to suspect that it's general relativity that breaks down at the event horizon, much the way Newton's laws faced a crisis when dealing with strong fields of gravity, such as those near the Sun or a massive neutron star. Einstein amended Newton, and now Einstein's theory will likely need revision to reveal a black hole's full character. Answers will arrive when physicists are able to join general relativity with quantum mechanics in that all-encompassing theory of quantum gravity.

Many have been trying for decades but are still far from success, yet there are hints on where it might lead. Many explorers into quantum gravity are coming to the conclusion that space-time itself—the central and core unit in Einstein's theory—may not be fundamental at all. It could be that space-time *emerges* (as physicists like to put it) out of other sorts of "bits," some kind of quantum grains that would be identified once a full theory of quantum gravity is in place. From this perspective, both space and time on the smallest of scales would have no meaning, just as a pointillist painting, built up from dabs of paint, cannot be fathomed close up. At that range, the painting looks like nothing more than a random array of dots. But as you move back, the dots begin to blend together and a recognizable picture slowly comes into focus. Likewise, space-time, the entity so familiar to us, might take form and reveal itself only when we scrutinize larger and larger scales. Space-time could be simply a matter of perception, present on the large scale but not on the smallest scale imaginable. You could think of space-time as "congealing" or "crystallizing" out of the chaotic quantum jumble that lies deep in the heart of the vacuum.

And it is at the black hole's event horizon where this new vision might be revealed. For decades, only astrophysicists or general relativists looking for the ultimate test case for Einstein's theory studied black holes. But now quantum physicists are mightily interested in black holes as well. They believe that clues to a unified theory of all the forces of nature may be found right there at the event horizon, that vital boundary where the microcosm of quantum mechanics directly meets the macrocosm of general relativity. Some recent models even suggest that any astronaut who dares to enter a black hole would not get the smooth entry past the event horizon, as described in chapter 10. Rather than an uneventful passage past the horizon and then a quick plummet to the singularity, they'd instead dramatically slam into a "firewall," where space and time are breaking up into its fundamental units. A singularity would no longer be featured. Saved from a plunge to a singularity, the astronauts would still be recycled into quantum bits.

But no one knows for sure. Physicists searching for the "theory of everything," such as string theory or loop quantum gravity, have no idea as yet what the final answer will reveal. Rather than a firewall, some other change (or no change at all) could arise as an event horizon is crossed. Toward the end of his life, John Wheeler held out the hope that the center of a black hole has a finite structure. He imagined that "the core of a black hole will prove to have some structure, albeit tiny beyond all imagining."

Explorers of this puzzle are much like the astrophysicists in the 1960s who were struggling to understand how traditional methods of power generation could possibly fuel a quasar's tremendous output, before realizing that the new kid on the block—the black hole that was hardly accepted as real—was operating as a dynamo, a real surprise to many.

Hawking in the 1970s began a conversation about the ultimate nature of black holes that continues to this day. Following in the footsteps of the pioneering quantum gravity theorists, he allowed more of his colleagues to see the profound connections between gravitation and quantum mechanics. Even though these two disparate laws of nature have yet to be officially joined, there are now fundamental signs that unity—the holy grail of physics—may well be achievable someday. And serving as physicists' prime guide in this aspiration is the black hole.

Epilogue

The Hanford Site, a prime repository for nuclear waste in the United States, sprawls over hundreds of square miles of scrub desert in south-central Washington State. There resides the Laser Interferometer Gravitational-Wave Observatory operated by the California Institute of Technology and the Massachusetts Institute of Technology. It's simply known as LIGO (pronounced LIE-go) for short.

Standing alone on the vast plain, a landscape long ago carved flat by the immense outflow of an ancient glacial lake, the complex resembles a modern art museum inexplicably placed in the middle of nowhere. An exact duplicate, painted in the same hues of cream, blue, and silver gray, can be found in the pine forests of Livingston Parish in Louisiana, outside Baton Rouge. Together, with similar observatories set up (or setting up) in Italy, Japan, and India, they form one of the most advanced astronomical tools of the twenty-first century.

The signals all these observatories are seeking are waves of gravitational radiation, or more simply gravity waves, as they are better known in the popular media. Einstein first wrote on their possible existence in 1916 and 1918, shortly after he had introduced his general theory of relativity. He had recognized that just as electromagnetic waves, such as radio waves, are generated when electrical charges travel up and down an antenna, waves of gravitational radiation are produced when masses move about. Electromagnetic waves—be they visible light, infrared, or radio waves—generally reveal a celestial object's physical condition, such as how hot it is, how old, and what it is made of. That's what standard astronomical observatories have been doing for decades. Gravity waves, on the other hand, will convey an entirely different story. They will tell us about the titanic movements of massive celestial objects.

Gravity waves are literally quakes in the very fabric of space-time, rumbles that emanate from the most violent events the universe has to offer—a once-blazing star burning out and going supernova, the dizzying spin of neutron stars, or the cagey dance of two black holes whirling around each other, approaching closer and closer until they collide in a spectacular merger. It is in this way that astronomers hope to obtain *direct* proof of a black hole's existence, the actual capture of a signal generated by the black hole itself. In this way, the black hole would obtain its ultimate affirmation.

The instrumentation for this endeavor, though, is vastly different from that of an ordinary telescope. There are no lenses to spy on the universe. Instead, long, wide tubes are set at right angles to each other. The LIGO tubes, for example, extend out into both the Washington and Louisiana countryside for two and a half miles (four kilometers), each forming a giant L in the landscape. Resembling oil pipelines, the tubes are as empty of air as the vacuum of space. And at each end mirrors have been suspended, with a laser beam continuously bouncing back and forth between them.

Gravity-wave observatories are set up in this way because of the unique effect a gravity wave has as it travels through and disturbs space-time. The wave *compresses* space in one direction—say, north and south—while simultaneously *expanding* it in the perpendicular direction—east and west. So, a gravity wave coming straight down on the L-shaped observatory would squeeze one of the arms, causing the mirrors at each end to draw closer together, while spreading the mirrors in the other arm farther apart. A millisecond later, as the gravity wave continues on its way, this effect would reverse, with the compressed arm expanding and the expanding arm contracting. The laser beams, continually measuring the distance between the mirrors, would take note of this cyclic change.

This is far trickier than it sounds. The gravity waves emitted by two black holes colliding are very powerful. Space is shaken, and shaken hard. Such a colossal collision would send out a spacequake that surges through the cosmos at the speed of light. But they wouldn't propagate in the manner a light wave travels through space. Rather, they would be an agitation of space itself. The waves would alternately compress and stretch the fabric of space-time as they travel along. Such waves would be deadly near the crash site. They could stretch a six-foot man to twelve feet and within a millisecond squeeze him down to three, before stretching him out once again. Any planets in the vicinity would be torn asunder. But those waves would get weaker and weaker as they travel outward, not unlike the ripples that die out after a rock is thrown into a pond. By the time those waves reach Earth, the expansion and contractions they create in space-time will be far smaller than the width of a proton particle.

To measure such a minuscule movement, gravity-wave astronomers have taken great pains to eliminate within their observatories as many local disturbances as possible, so that a passing truck or seismic tremor could be ruled out. And as the data streams in, it is continually compared to various "templates," theoretical predictions of what a gravity-wave signal might look like from differing events. Simultaneous detections of a signal by separate observatories, located half a continent or more apart, would be the best confirmation.

Neutron-star collisions may be their bread-and-butter sources. The observatories are set up to register the final minutes of a neutron-star binary, as the two city-sized balls of dense matter spiral into one another. LIGO will be most receptive to signal frequencies from 100 to 3,000 cycles per second, which is coincidentally the same frequencies our ears pick up as sound. So once the wave is electronically recorded, you could actually listen to it. Gravity-wave observatories will be adding sound to our cosmic senses. The neutron-star clash would start off as a whine and then rapidly rise in pitch, like the sound of a swiftly approaching ambulance siren.

The biggest prize of all, though, will be two black holes colliding. As the twirling holes are about to meet, spiraling inward faster and faster at speeds close to that of light, it's predicted the whine will turn into a quick *chirp,* a birdlike trill that races up the scale in a matter of seconds. A cymbal-like crash, a mere millisecond in length, would herald the final collision and merger. The two black holes become one, followed by a ring-down, akin to the diminishing tone of a struck gong, as the new and bigger hole wobbles a bit and then settles down.

There are ways to detect gravity waves indirectly. Teams of radio astronomers operating very sensitive detectors at the South Pole are on the lookout for the distinct imprint of gravity waves upon the Big Bang's now-weak afterglow—a wash of radio waves known as the cosmic microwave background. Gravity waves can affect the microwaves in an unmistakable way, imposing a faint spiraling pattern into the polarization of the signal. The gravity waves were born as quantum fluctuations in the newborn force of gravity itself, whizzing through the tiny speck of a universe. These waves were fueled and blown up in size as the cosmos underwent a brief, accelerated spurt of expansion called inflation in the first trillionth of a trillionth of a trillionth of a second of our cosmic beginnings, when afterward the cosmos settled down into a slower expansion. By stretching and squeezing space-time, the primordial gravity waves can imprint a slight swirling pattern on the remnant radiation that had become "polarized" (the electric fields of the light waves oscillating back and forth in one preferred direction) as the light began to freely traverse the cosmos. As they rippled space-time, the waves gives the light a little kick that causes its orientation to curl. Dust in our galaxy can lead to the same effect, so any signal must be closely examined to confirm that the primordial universe was the source.

If such a signal is verified, it would be the first detection of Hawking radiation, although not in the context of black-hole event horizons. The observable universe was so tiny at first that it also had a "horizon," which emits radiation just as hypothesized for a black hole. In this case, the radiation is in the form of gravitons, those quantized particles of gravity that grow into the gravity waves that stretched and squeezed the primordial radiative soup. If the signature of Hawking radiation is assuredly found within the Big Bang, it makes it highly likely that such radiation is emitted by black holes as well. This would open up an entirely new arena for astronomy and cosmology, one that began years earlier when Jacob Bekenstein and Stephen Hawking began thinking outside the box.

There's more definitive proof of gravity waves' effects closer in. Two neutron stars in our galaxy, located some twenty-one thousand light-years distant, are rapidly orbiting each other and also drawing closer and closer together. The rate of their orbital decay—the orbit shrinking by about 11.5 feet (3.5 meters) each year—is just the change physicists expect if this binary pair is losing orbital energy in the form of gravitational waves. The amount of energy the waves carry off matches, with exquisite precision, just what general relativity predicts. Astronomers Joseph Taylor and Russell Hulse won the Nobel Prize in 1993 for this discovery. The gravity waves being emitted by this system are currently too weak to be recorded by the earthbound observatories, but the gravitational ripples will be far more powerful once the two stars finally merge about three hundred million years from now.

But there are lots of other gravity-wave sources that are currently detectable, including supernova explosions, black-hole mergers, and neutron-star collisions that regularly occur throughout the cosmos. Once the observatories are fully up and running, sensitive enough to pick up waves originating from up to billions of light-years away, scientists hope to see some kind of event daily. Even more sources could be recorded if the technology is taken up to space, far from ground-based disruptions, endeavors that are on the drawing board.

Firm detection of a gravity wave is such a high priority in relativistic astrophysics that scientists are not relying on one method alone. There's another clever scheme based on well-studied astronomical objects—pulsars, the most exquisite timepieces in the universe. By closely monitoring the pulses arriving from an array of particularly fast pulsars, situated around the sky, astronomers are on the lookout for slight changes in the pulsing due to a gravity wave passing between the pulsar and the earthbound detector. No matter which way the gravity wave from a black hole is detected, such sightings would provide the final, undeniable proof that black holes are real, which would be a historic moment for astronomers, who for so long denied their existence.

A rare polar express swept over the Dallas area in December 2013, heavily icing both the airport and the roads. It was a freezing start to the 27th Texas Symposium on Relativistic Astrophysics, its fiftieth anniversary. Held nearly every two years since 1963, the conference has been located in cities around the globe, from Munich and Melbourne to Jerusalem and Vancouver. Yet, wherever it is held, the meeting still retains its Texas name to honor its origin.

The first symposium was primarily focused on quasars, with the phrase "relativistic astrophysics" newly introduced. Over the next fifty years the list of topics discussed at the symposium expanded like a raging wildfire; there are now sessions on the inflationary universe, gravitational waves, searches for dark matter, gamma ray bursts, and the cosmic microwave background. Some of these subjects weren't even imagined half a century ago, when pulsars hadn't yet been discovered. "We didn't know that neutron stars would come equipped with a handle and a bell," as one wit quipped. Now more than 2,300 pulsars have been cataloged within our galaxy.

As for black holes, no longer are eyebrows raised at the sound of their name. In fact, the 2013 Texas Symposium served up a succulent banquet of talks on the subject; researchers reported on the origin of supermassive black holes, gamma-ray bursts from newly formed holes, black-hole mergers, magnetized black holes, jets shooting out of black holes, and new searches for these collapsed objects. They are as readily discussed at current astronomy conferences as a galaxy, nebula, or star. The stellar-sized holes are just another possible endpoint (albeit rarer) in the lifetime of a star. It's estimated about one star in a thousand ends its life hidden behind an event horizon, with one hundred million of them residing in the Milky Way alone. A new one is being born somewhere in the cosmos with each tick of the clock. And the supermassive ones, grandly residing in the hearts of most galaxies, are now standard equipment in a galaxy's very structure.

John Wheeler once remarked that he never read science fiction. "All the science fiction I need is right out there in front of us," he said. He was absolutely right. Black holes, long considered so fanciful, have now transformed into some of the most wondrous and necessary denizens of the cosmos. Once scorned but now accepted, the black hole is commencing a brand new chapter of its life.

P.S. And that new chapter began with an ever-so-short but enchanting "song" on 14 September 2015. In the wee hours of the morning that day, both LIGO observatories simultaneously registered a passing gravity wave, the first ever recorded. The signal lasted just a few tenths of a second. The wave started at 30 cycles per second—a deep bass—and swiftly rose in a crescendo to a low A-sharp or B note, nearly 250 cycles per second. It was music to LIGO scientists' ears, a symphony they've been waiting decades to hear.

From the signal's properties, the LIGO scientists knew they had captured the last gasp as two massive black holes, wildly spiraling around one another, finally merged in a fateful embrace. Their relativistic equations told them that the two holes collided 1.6 billion years ago, when blue-green algae were the highest form of life on Earth. The momentous clash shook space-time hard, generating a gravity wave that traveled a distance of 1.6 billion light-years until it hit the shores of Earth.

This was a landmark moment. No longer is the belief in black holes based on theoretical models or supposition. The gravity wave captured that September night was a direct and collective shout from the two black holes themselves. Here we are, they were saying, here we are.

Timeline

1687 Sir Isaac Newton publishes his revolutionary law of gravity in the *Principia.*

1758 A comet predicted by Edmond Halley to return in 1758 arrives on schedule, a victory for Newton's law of gravity.

1783 John Michell in Great Britain introduces the Newtonian version of a black hole. He calculates the mass at which a star would be so heavy that light cannot escape from it, rendering it invisible.

1796 Following similar reasoning as Michell, Pierre-Simon de Laplace in France independently suggests the existence of *corps obscurs,* or hidden stellar bodies, in the heavens.

1862 Alvan Graham Clark in Massachusetts discovers that the bright star Sirius has a faint companion. But astronomers were later puzzled how it could be so dim and yet still weigh as much as the Sun.

1905 Albert Einstein publishes what came to be known as his special theory of relativity, which abolishes Newtonian notions of absolute space and time.

1907 Mathematician Hermann Minkowski demonstrates that Einstein, with special relativity, has turned time into just another dimension, leading to the single, absolute entity of space-time.

1915 Introducing his general theory of relativity, Einstein successfully broadens relativity to handle other types of motion, specifically gravity. Gravity is now seen as arising from masses indenting the flexible mat of space-time, with objects moving along the curvatures.

1916 The German astronomer Karl Schwarzschild publishes the first full solution to the equations of general relativity. The result leads to the appearance of the Schwarzschild sphere, which surrounds a mass concentrated at a point in its center. Space and time appear to stop at the sphere's surface. It is a version of what we today call the black hole, this one uncharged and nonspinning. Some assume it is an artifact of the coordinates being used; others are sure stars could never shrink to such a state.

Estonian Ernst Öpik and later Britisher Arthur Eddington calculate that the solar-mass companion of Sirius must be little larger than the Earth, which explained its faintness. Such stars came to be called "white dwarfs."

1919 British solar eclipse expeditions to West Africa and Brazil verify that starlight indeed bends its path as it passes close to the Sun, following the indentation the Sun makes in space-time. General relativity is triumphant.

1926 British theorist Ralph Fowler uses the newly established rules of quantum mechanics to explain how the mass of the Sun, crushed to an Earth-sized space, can remain stable as a white dwarf star.

1930 On a voyage from India to Great Britain, Subrahmanyan Chandrasekhar discovers that there is a maximum limit to the mass of a white dwarf star. He doesn't know what happens to the star should it pass that threshold.

1931 The Soviet theorist Lev Landau calculates a star could collapse to a point if heavy enough, but deeming such a result "ridiculous," he suggests that the stellar core instead forms "one gigantic nucleus."

1932 James Chadwick in Great Britain discovers the neutron.

Bell Telephone physicist Karl Jansky discovers radio waves emanating from the center of the Milky Way galaxy. Radio astronomy begins.

1933 At a meeting of the American Physical Society, Fritz Zwicky and Walter Baade suggest that a tiny neutron star forms in the midst of a stellar explosion, a supernova. Astronomers consider the idea farfetched.

1935 At a meeting of the Royal Astronomical Society, Arthur Eddington infamously challenges Chandrasekhar's conclusion that a white dwarf star would suddenly shrink if its density passed a maximum limit.

1939   J. Robert Oppenheimer and George Volkoff are the first to examine the physics of a neutron star and discover that neutron stars, just like white dwarfs, have a maximum limit to their mass.

Oppenheimer and Hartland Snyder publish the first modern description of a black hole. They call it "continued gravitational contraction." Oppenheimer then abandons this line of work. Interest in general relativity continues to plummet within the physics community.

Einstein publishes his "worst scientific paper," an attempt to prove that stars could never totally collapse to a point (or singularity).

1948   American financier Roger Babson founds the Gravity Research Foundation to renew interest in gravitational studies (so that anti-gravity devices might be developed one day). Its funding spurs new interest in general relativity.

1952–53   Princeton physicist John Archibald Wheeler teaches the first courses on special and general relativity ever offered by his physics department. He hopes to find a physical reason that prevents a star from collapsing to a singularity, as Oppenheimer and Snyder had suggested.

1955   Einstein dies, believing his colleagues regarded him as an "old fool" and his greatest work—general relativity—in the shadows of physics research.

1957   A new institute for gravitational studies at the University of North Carolina holds a conference at Chapel Hill on the role of gravitation in physics. The meeting proves to be a landmark in reenergizing the field.

At an international conference John Wheeler and two student collaborators attempt to demonstrate how an imploding star could save itself from collapse to a singularity. Oppenheimer, in the audience, politely disagrees.

1958   David Finkelstein develops a new reference frame for general relativity that makes it easier to comprehend the physics of a black hole. It allows physicists to picture how a collapsing star appears like a "frozen star" from afar, yet still fully implodes from the standpoint of the hole. Martin Kruskal did this earlier, but his work wasn't published until 1960.

~ 1960 At a colloquium at the Institute for Advanced Study, Princeton physicist Robert Dicke jokingly compares the complete collapse of a star, where gravity is so strong that nothing can escape, to the "Black Hole of Calcutta." Physicist Hong-Yee Chiu is in the audience.

1962 Using the new mathematical tools developed by Finkelstein and Kruskal, Princeton undergraduate David Beckedorff, working with Charles Misner, arrives at a more detailed description of the space outside a black hole's event horizon. It was the first depiction of a black hole as a real object.

The field of X-ray astronomy is initiated when a rocketborne X-ray detector discovers the first cosmic X-ray source, Sco X-1, later revealed to be a neutron star in a binary star system.

Early 1960s Computer simulations carried out at the Livermore National Laboratory in California demonstrate that a star of sufficient mass at the end of its life will collapse to a black hole. Similar results are obtained by Soviet physicists. Convinced by these findings, as well as Beckedorff's results, Wheeler completely reverses his opinion and comes to champion black holes. The Soviets hardly doubted their existence.

1963 A radio star known as 3C 273 is revealed to be the superluminous nucleus of a galaxy some two billion light-years distant. Such objects are soon dubbed "quasars."

Roy Kerr successfully solves a decades-old challenge in general relativity; he devises a solution to Einstein's equations that fully models the gravitational field of a star that is rotating.

The First Texas Symposium on Relativistic Astrophysics is held in Dallas to try to figure out the source of a quasar's astounding power. This conference was the first notable attempt to link general relativity with astrophysical concerns.

1964 The term *black hole* is seen in print for the first time in the 18 January 1964 issue of *Science News Letter,* which was reporting on an astronomy session on degenerate stars at the annual meeting of the American Association for the Advancement of Science (AAAS). Borrowing the term from Robert Dicke, Hong-Yee Chiu, chair of the session, suggested that space was peppered with black holes.

Soviet physicists Yakov Zel'dovich and Igor Novikov, and independently Cornell University physicist Edwin Salpeter, suggest that great energies can be released as matter is drawn to a supermassive collapsed object, forming an accretion disk around it, which could explain a quasar's long-lasting power.

1965 British physicist Roger Penrose proves theoretically that gravitational collapse always results in the formation of a singularity inside a black hole.

1967 John Wheeler uses the phrase "black hole" to describe his gravitationally collapsed object during a keynote address at the annual meeting of the AAAS. Once his lecture is published in 1968, the scientific community begins to embrace the term as the object's official name.

British astronomer Jocelyn Bell discovers pulsars, later understood to be spinning neutron stars. The finding convinces many that black holes, too, might exist.

1969 Roger Penrose shows how enormous energies can be extracted from a black hole's rapid spin.

1971 Based on data from the X-ray satellite *Uhuru,* an atypical radio source known as Cygnus X-1 is tentatively identified as a black hole, the first to be discovered in space.

1973 Jacob Bekenstein publishes that the area of a black hole's event horizon is a direct measure of the hole's entropy.

1974 Attempting to prove Bekenstein wrong, Stephen Hawking instead proves that a black hole can gradually evaporate as it releases radiation ("Hawking radiation") over time. His finding is a historic link between general relativity and quantum mechanics.

Kip Thorne and Stephen Hawking make a bet on whether Cygnus X-1 is truly a black hole. Thorne is for, Hawking against.

1977 Roger Blandford and Roman Znajek develop their model for extracting energy out of a spinning black hole.

1990 Stephen Hawking concedes to Kip Thorne and agrees that Cygnus X-1 is a black hole.

1999 Construction is completed on the Laser Interferometer Gravitational-Wave Observatory, with one installation in Washington State and another in Louisiana. Operation of the observatory begins in 2001; more advanced detectors are scheduled to be operating in 2015. A gravity-wave signal would provide the first direct proof of a black hole's existence.

2013 The Texas Symposium, held once again in Dallas, celebrates its fiftieth anniversary. Black holes are now fully accepted. Talks are given on black-hole mergers, their magnetization, energy production, and the gamma-ray bursts they can emit upon their birth.

2015 On 14 September the LIGO observatories recorded the first gravity wave ever detected. The gravity-wave signal was emitted by a pair of black holes as they merged and combined to form one massive black hole located 1.6 billion light-years away. This gravity wave provided the first direct evidence that black holes exist.

Notes

Abbreviations

APS, American Philosophical Society Library, Philadelphia
AIP, American Institute of Physics, Niels Bohr Library and Archives, College Park, Maryland

Preface

"Like unicorns and gargoyles": Thorne, *Black Holes and Time Warps*, 23.

"Nearly everyone understands": Wheeler, *Cosmic Catastrophes*, 176.

"All truth passes through three stages": The origin of this quotation is often attributed to the nineteenth-century philosopher Arthur Schopenhauer. In the preface to his *Die Welt als Wille und Vorstellung* (The world as will and representation, 1818), Schopenhauer wrote, "Der Wahrheit ist allerzeit nur ein kurzes Siegesfest beschieden, zwischen den beiden langen Zeiträumen, wo sie als Paradox verdammt und als Trivial gering geschätzt wird," which translates as, "To truth only a brief celebration of victory is allowed between the two long periods during which it is condemned as paradoxical, or disparaged as trivial." Many others proceeded to devise variations of this statement. See Shallit, "Science, Pseudoscience," 2.

"Einstein's predictions": Chandrasekhar, "Theory of Relativity," 249.

"work of art": Born, "Physics and Relativity," 253.

energy of a trillion suns: Begelman and Rees, *Gravity's Fatal Attraction*, III.

"Beauty is the splendor of truth": Wali, "Chandra," 13. See also "Subramanyan Chandrasekhar—Nobel Lecture: On Stars, Their Evolution and Their Stability" at http://www.nobelprize.org/nobel_prizes/physics/laureates/1983/chandrasekhar-lecture.html.

"We will first understand": Wheeler and Ford, *Geons, Black Holes*, 5.

Chapter 1: It Is Therefore Possible That the Largest Luminous Bodies in the Universe May Be Invisible

everything fell toward it: discussed in Aristotle's *De caelo* (On the heavens), his cosmological treatise written in the fourth century BCE.

Sun was now at center stage: Copernicus set out his theory in *De Evolutionibus Orbium Coelestium* (On the revolutions of the heavenly spheres), published in 1543, the year of his death.

Earth was a giant magnet: see William Gilbert, *De Magnete* (On the magnet, 1600).

threads of magnetic force: discussed in Johannes Kepler's *Epitome Astronomiae Copernicanae* (Epitome of Copernican astronomy, 1618–1621).

vortices of *aether:* described in René Descartes, *Le Monde* (The world), written between 1629 and 1633 and not published in its entirety until 1677).

"He hesitated and floundered": Westfall, *Never at Rest,* 155.

developed an intriguing set of conjectures: Hooke's paper, titled *Attempt to Prove the Motion of the Earth,* was published in 1674. He republished it in 1679 in his *Lectiones Cutlerianae.*

"Circle, Ellipsis": Westfall, *Never at Rest,* 382.

"I am . . . shy": Brewster, *Memoirs of Newton,* 193.

"An ellipsis": Westfall, *Never at Rest,* 403.

"ecstasy, total surrender": Westfall, *Never at Rest,* 103.

Kepler had revealed: Kepler, *Astronomia Nova* (The new astronomy, 1609).

"For nature is simple": Newton, *Principia,* 794.

"I have not as yet": Newton, *Principia,* 943.

comet would return: in Halley, *Astronomy of Comets.*

"the most inventive": McCormmach, "Michell and Cavendish," 127.

"father of modern seismology": Hardin, "Scientific Work," 30.

Michell was able to compute: Jungnickel and McCormmach, *Cavendish,* 185.

Cavendish ultimately obtained the torsion balance: Crossley, "Mystery," 62.

burying original insights: Crossley, "Mystery," 66.

"short Man": Montgomery, Orchiston, and Whittingham, "Michell, Laplace," 90.

"comet of the first magnitude": Crossley, "Mystery," 69.

"continued to indulge": Montgomery, Orchiston, and Whittingham, "Michell, Laplace," 90.

"odds against the contrary opinion": Michell, "Inquiry into the Probable Parallax," 249.

"arguably the most innovative": Montgomery, Orchiston, and Whittingham, "Michell, Laplace," 91.

"those fires": Michell, "Inquiry into the Probable Parallax," 238.

"a most valuable present": Michell, "Means of discovering the Distance," 36.

"On the Means": Michell, "Means of discovering the Distance," 36.

over a succession of meetings: Jungnickel and McCormmach, *Cavendish,* 344–45.

Michell was devoted to the society: Jungnickel and McCormmach, *Cavendish,* 565, n. 7.

weren't detecting: McCormmach, "Michell and Cavendish," 149.

some historians have speculated: Jungnickel and McCormmach, *Cavendish,* 564.

pull should also affect light: Michell, "Means of discovering the Distance," 36–37.

"all light": Michell, "Means of discovering the Distance," 42.

"A luminous star": Laplace, *System of the World,* 367.

appeal from a dogged colleague: Montgomery, Orchiston, and Whittingham, "Michell, Laplace," 93.

invisible-star speculation: Gillispie, *Laplace,* 175.

revolved around a luminous star: Michell, "Means of Discovering the Distance," 50.

Chapter 2: Newton, Forgive Me

"velocity is so nearly that of light": Maxwell, "Dynamical Theory," 466.

"the only occupation": Maxwell, "Introductory Lecture," 244.

"not correct": Einstein, *Collected Papers,* 1:131.

"no such thing": Einstein, "Autobiographical Notes," 53.

"introduce another postulate": as translated in Lorentz et al., *Principle of Relativity,* 38. The original paper is Einstein, "Elektrodynamik beweger Körper."

"superfluous": Lorentz et al., *Principle of Relativity,* 38.

"an 'absolutely stationary space'": Lorentz et al., *Principle of Relativity,* 38.

"for me": Born, "Physics and Relativity," 250.

"space by itself": Minkowski, "Space and Time," 75. This was originally presented as an address to the Eightieth Assembly of German Natural Scientists and Physicians, Cologne, Germany, 21 September 1908.

"banal": Fölsing, *Einstein,* 245.

"superfluous learnedness": Pais, *"Subtle Is the Lord,"* 152.

"child's play": Einstein, *Collected Papers,* 5:324.

happiest times in his life: Stachel, *Einstein from "B" to "Z,"* 5.

"like a child": Eisenstaedt, *Curious History of Relativity,* 67.

"stuck in its diapers": Fölsing, *Einstein,* 245. This quotation comes from a book that Einstein wrote for the public in 1917 on special and general relativity titled *Über die spezielle und allgemeine Relativitätstheorie, gemeinverständlich* (On the special and general theory of relativity, generally comprehensible). Others have translated the quotation into English as general relativity "would perhaps have got no farther than its long clothes." I chose Fölsing's translation.

"Grossman, you must help me": Pais, *"Subtle Is the Lord,"* 212.

"I was beside myself": Hoffmann, *Einstein,* 125.

"boldest dreams": Einstein, *Collected Papers,* 8:160.

"Spacetime tells matter": Wheeler and Ford, *Geons, Black Holes,* 235.

"Newton, forgive me": Einstein, "Autobiographical Notes," 31.

Einstein in 1911 had suggested a specific test: Einstein, "Influence of Gravity."

"We have no time to snatch a glance at [the Sun]": Eddington, *Stars and Atoms,* 115.

unscientifically rooting for Einstein: Eddington, *Stars and Atoms,* 116.

"LIGHTS ALL ASKEW": *New York Times,* 10 November 1919, 17.

"Like the man in the fairy tale": Einstein, *Collected Papers,* 10:265.

Chapter 3: One Would Then Find Oneself . . . in a Geometrical Fairyland

in less than a month: Schwarzschild wrote Einstein from the Russian front with his solution on 22 December 1915. See Einstein, *Collected Papers,* 8:163–65.

"Mr. Einstein's result": *Dictionary of Scientific Biography,* s.v. "Schwarzschild, Karl." He wrote those words in his paper "On the Gravitational Field of a Mass Point According to Einstein's Theory," first published in *Sitzungsberichte der Königlich Preussischen Akademie der Wissenschaften zu Berlin, Phys.-Math. Klasse* (1916): 189–96. In another English translation (Schwarzschild, "Gravitational Field of a Mass Point," 952), the phrase is given as "let Mr. Einstein's result shine with increased clearness."

"We can wonder": Schwarzschild, "Ueber das zulässige Krümmungsmaass des Raumes," 337. Translated from the German, "Man kann die Vorstellungen bis ins Einzelnste ausbilden, wie die Welt in einem sphärischen oder pseudosphärischen [Geometrie]. . . . Man befindet sich da—wenn man will—in einem geometrischen Märchenland, aber das Schöne an diesem Märchen ist, dass man nicht weiss, ob es nicht am Ende doch Wirklichkeit ist." The translation used is from Chandrasekhar, *Truth and Beauty,* 146.

avidly followed Einstein's progress: AIP, Spencer Weart interview with Martin Schwarzschild, 10 March 1977.

wanted to remove all doubt: Schemmel, "Astronomical Road," 465.

"The consequences": Sampson, "Principles of Relativity and Equivalence," 155.

"they got stuck": Eisenstaedt, *Curious History of Relativity,* 266.

"The mass would produce": Eddington, *Internal Constitution of the Stars,* 6.

"They realized": Eisenstaedt, *Curious History of Relativity,* 264.

best way to describe this unusual place: Eisenstaedt, *Curious History of Relativity,* 307–8.

the magic sphere would stretch: Wheeler, *Cosmic Catastrophes,* 179.

He figured that Schwarzschild's novel entity: Earman and Eisenstaedt, "Einstein and Singularities," 186. Einstein mentioned this specifically in *Meaning of Relativity,* 3rd ed., 124, and in earlier editions of the book.

Many viewed Schwarzschild's sphere: see Piaggio and Critchlow, "Supposed Relativity Method."

worked out a little calculation: Eisenstaedt, *Curious History of Relativity,* 261.

densities could ever be greater: see Jeffreys, "Compressibility of Dwarf Stars."

"clearly not physically meaningful": Schwarzschild, "Gravitational Field of a Sphere," 434. See http://cds.cern.ch/record/412373/files/9912033.pdf.

had been in the audience: Schemmel, "Astronomical Road," 464.

completed two papers: Schwarzschild, "Gravitational Field of a Mass Point," and "Gravitational Field of a Sphere."

"As you see, the war": Einstein, *Collected Papers,* 8:164.

always ready for a good beer: Weart interview with Schwarzschild.

"I would not have expected": Einstein, *Collected Papers,* 8:175.

"there will come a time": Anderson, "Advance of the Perihelion," 627.

"concentration to that extent": Lodge, "Supposed Weight and Ultimate Fate of Radiation," 551.

"A stellar system": Lodge, "Supposed Weight and Ultimate Fate of Radiation," 551.

Chapter 4: There Should Be a Law of Nature to Prevent a Star from Behaving in This Absurd Way!

slight but distinct wobble: Bessel, "Variations of Proper Motions."

completed one orbit: Bessel, "Variations of Proper Motions," 139.

"The subject": Bessel, "Variations of Proper Motions," 136.

closest to gleaming Sirius: Holberg and Wesemael, "Discovery of the Companion of Sirius," 167.

testing the optics: Holberg and Wesemael, "Discovery of the Companion of Sirius," 162.

"there might have been a [prearranged] connection": Welther, "Discovery of Sirius B," 34.

"It remains": Bond, "Companion of Sirius," 286–87. See also Holberg and Wesemael, "Discovery of the Companion of Sirius," 165.

Lalande Prize: Holberg and Wesemael, "Discovery of the Companion of Sirius," 170–71.

sunlike star cooling off: DeVorkin, "Hertzsprung-Russell Diagram," 32.

companion known since 1783: It was discovered by William Herschel, during his double-star searches, on 31 January 1783. See Herschel, "Catalogue of Double Stars," 73.

confirmed the spectrum: Adams, "A-Type Star."

"I was flabbergasted": Philip and DeVorkin, "In Memory of Henry Norris Russell," 90.

Adams determined: Adams, "Spectrum of the Companion of Sirius."

Soon theorists: see Öpik, "Densities of Visual Binary Stars," and Eddington, "Relation Between the Masses."

"an impossible result": Öpik, "Densities of Visual Binary Stars," 302.

"The message": Eddington, *Stars and Atoms,* 50.

Ralph Fowler: Fowler, "On Dense Matter."

"It is something": AIP, Spencer Weart interview with Subrahmanyan Chandrasekhar, 17 May 1977. See http://www.aip.org/history/ohilist/4551_1.html.

result published as a brief 1931 paper: Chandrasekhar, "Maximum Mass of Ideal White Dwarfs."

"I am sorry": Weart interview with Chandrasekhar.

a stronger (and stranger) conclusion: Wali, "Chandra's Work in Historical Context."

"inconceivable": Chandrasekhar, "Highly Collapsed Configurations," 463.

"there exists in the whole quantum theory": Landau, "Theory of Stars," 287.

"ridiculous" result: Landau, "Theory of Stars," 287.

a thought that the Danish atomic physicist Niels Bohr: Hufbauer, "Landau's Youthful Sallies," 340.

The stellar core was a "pathological" region . . . "one gigantic nucleus": Landau, "Theory of Stars," 287–88.

"We may conclude": Chandrasekhar, "Remarks on the State of Matter," 327.

The top Britishers in stellar physics—Eddington, James Jeans, and Edward Milne—were too busy arguing: Miller, *Empire of the Stars*, 81–82.

"It is necessary": Chandrasekhar, "Stellar Configurations," 377.

"but I kept away from it": Weart interview with Chandrasekhar.

"It seemed to me": Weart interview with Chandrasekhar.

He decided to take on the challenge: Thorne, *Black Holes and Time Warps*, 153.

This time he had an exact solution: Wali, *Chandra*, 124.

"Involved in the puzzles": Miller, *Empire of the Stars*, 103.

"when the central density": Chandrasekhar, "Highly Collapsed Configurations (Second Paper)," 207.

"there should be a law of Nature": "Discussion of Papers," 38.

needed calculator: Miller, *Empire of the Stars*, 11.

"I do not know whether I shall escape": "Discussion of Papers," 38–39.

"an unholy alliance": Eddington, "Relativistic Degeneracy," 195.

Chandra's defense: Thorne, *Black Holes and Time Warps*, 162.

"Stoner's findings": Nauenberg, "Edmund C. Stoner," 301.

Eddington was mainly bullying Chandra: Miller, *Empire of the Stars*, 109.

"stellar buffoonery": Chandrasekhar, *Truth and Beauty*, 132.

"debates were a sport": e-mail communication with Werner Israel, 2 December 2013.

"a sort of Don Quixote": Wali, *Chandra*, 145.

"Chandrasekhar limit": Starting in the early 1960s, papers in the astrophysical journals began referring to "Chandrasekhar's limiting mass."

"I had to make up my mind": Wali, *Chandra*, 146.

"In 1935 the astronomical community": Hawking and Israel, *Three Hundred Years of Gravitation*, 222.

Chandra admitted: Chandrasekhar, "Black Hole in Astrophysics," 5.

Chapter 5: I'll Show Those Bastards

on the night of 29 August 1975: International Astronomical Union Circular No. 2826, 1975; Mobberley, *Cataclysmic Cosmic Events*, 52; Shipman, *Restless Universe*, 308; "Fascination with Celestial Events."

Hipparchus: Russell, "Address," 2.

six thousand light-years: The distance is still inexact. See Wade et al., "Sharpened Hα + [N II] Image," 1738.

intense gravitational field: Comins and Kaufmann, *Discovering the Universe,* 222.

"There is something uncanny": Russell, "Address," 2.

"phenomena of a different order": Russell, "Address," 4.

"giant novae": Osterbrock, *Walter Baade,* 57.

"exceptional novae": see, e.g., Hubble, "Spiral Nebula," 127.

"Hauptnovae": Osterbrock, *Walter Baade,* 57.

Baade had a hip defect: Osterbrock, *Walter Baade,* 3.

"saw the mysteries": Robinson, Schild, and Schücking, *Quasi-Stellar Sources,* xi.

regularly photographed: Osterbrock, *Walter Baade,* 8.

exceptional novae: Baade and Zwicky, "Super-Novae." Astronomer Donald Osterbrock believes this paper was largely the work of Baade. See Osterbrock, *Walter Baade,* 58.

Knut Lundmark: Lundmark, "Pre-Tychonic Novae."

astronomy his "hobby": Zwicky, *Morphological Astronomy,* 11.

existence of cosmic "dark matter": Zwicky, "Die Rotverschiebung," 122.

"Zwicky was one of those people": Bartusiak, *Through a Universe Darkly,* 196.

James Chadwick had bombarded: Chadwick, "Possible Existence of a Neutron."

"Such a star": Baade and Zwicky, "Cosmic Rays from Super-Novae," 263. While Baade carried out the astronomical observations on supernovae, it's believed Zwicky was largely responsible for the speculative theoretical ideas. See Osterbrock, *Walter Baade,* 58–59.

packing the mass of a high-rise building: Wheeler, *Cosmic Catastrophes,* 36–37.

bathroom sink: McClintock, "Do Black Holes Exist?," 30.

"may be the origin": Chandrasekhar, "White Dwarfs," 245.

Chapter 6: Only Its Gravitational Field Persists

noted for their creative insight: Hufbauer, "Landau's Youthful Sallies," 344.

astute practitioner: Hufbauer, "Landau's Youthful Sallies," 345.

"neutronic core": Landau, "Origin of Stellar Energy," 334.

liberating enough energy: Landau, "Origin of Stellar Energy," 334.

"analogous to the conditions": Gamow, *Structure of Atomic Nuclei,* 235.

"be quite enough": Gamow, *Structure of Atomic Nuclei,* 238.

published the paper: Landau, "Origin of Stellar Energy."

"we can regard": Landau, "Origin of Stellar Energy," 334.

"bold idea": Thorne, *Black Holes and Time Warps*, 186.

anti-Stalinist leaflet: Hufbauer, "Landau's Youthful Sallies," 352.

Soviet physicist Pyotr Kapitsa: Thorne, *Black Holes and Time Warps*, 186.

urged his colleague J. Robert Oppenheimer: Cassidy, *Oppenheimer*, 175.

"Oppenheimer was already aware": Hufbauer, "Path to Black Holes," 39.

could not possibly harbor neutronic cores: Oppenheimer and Serber, "Stability of Stellar Neutron Cores." In a footnote, the authors thank Bethe for their discussions on these questions.

whether Landau's neutron cores: Hufbauer, "Landau's Youthful Sallies," 352.

driven to private school: Cassidy, *Oppenheimer*, 16.

filled the family's Manhattan apartment: Cassidy, *Oppenheimer*, 17.

"What really made": Cassidy, *Oppenheimer*, 135.

"something like a god": Cassidy, *Oppenheimer*, 154.

work with Max Born: Ferreira, *Perfect Theory*, 58.

Landau was convinced: Hufbauer, "Landau's Youthful Sallies," 341.

helped organize a symposium: Cassidy, *Oppenheimer*, 174.

"Tolman consulting": Hufbauer, "Path to Black Holes," 41–42.

Volkoff labored over a calculating machine: Hufbauer, "Path to Black Holes," 42.

eight-page paper: Oppenheimer and Volkoff, "Massive Neutron Cores."

"tour de force": Thorne, *Black Holes and Time Warps*, 207.

"the question of what happens": Oppenheimer and Volkoff, "Massive Neutron Cores," 380.

It was still possible: Oppenheimer and Volkoff, "Massive Neutron Cores," 381.

"Oppie was extremely cultured": Thorne, *Black Holes and Time Warps*, 212.

"very odd": Smith and Weiner, *Robert Oppenheimer*, 208–9.

"Only its gravitational field persists": Oppenheimer and Snyder, "Continued Gravitational Contraction," 456.

"most daring and uncannily prophetic": Hawking and Israel, *Three Hundred Years of Gravitation*, 226–27.

"golden list": Hufbauer, "Landau's Youthful Sallies," 353.

"graduate students or third-rate hacks": Dyson, "Chandrasekhar's Role," 47. Oppenheimer once told physicist Hong-Yee Chiu that his paper on gravitationally collapsed objects "was just an exercise for his students to work on as a Ph.D. thesis." E-mail communication with Chiu, 3 January 2014.

"most important contribution": Dyson, "Chandrasekhar's Role," 46.

Even Einstein: Einstein, "Stationary System with Spherical Symmetry."

Einstein stacked the deck: Earman and Eisenstaedt, "Einstein and Singularities," 225.

"One could not be sure": Einstein, "Stationary System with Spherical Symmetry," 922.

"great number": Einstein, "Stationary System with Spherical Symmetry," 922–23.

"strong candidate": Earman and Eisenstaedt, "Einstein and Singularities," 230.

"was so firmly convinced": Thorne, *Black Holes and Time Warps*, 137.

"There is a curious parallel": Thorne, *Black Holes and Time Warps*, 139.

Theorists revered: Chandrasekhar, *Truth and Beauty*, 69.

"Jewish physics": Beyerchen, *Scientists Under Hitler*, 132–33.

universities around the world rarely offered: This is not unprecedented. In 1904, some four decades after Maxwell introduced his laws of electromagnetism, Lord Kelvin wrote, "The so-called 'electromagnetic theory of light' has not helped us hitherto. . . . It seems to me that it is rather a backward step." Kelvin, Baltimore Lectures, vii, 9.

"It was despised": Eisenstaedt, *Curious History of Relativity*, 242.

"with some tricky concepts": Eisenstaedt, *Curious History of Relativity*, 234.

"probably also our brain processes": Eisenstaedt, *Curious History of Relativity*, 234.

"Professor Eddington": Chandrasekhar, *Truth and Beauty*, 117.

Less than 1 percent: Eisenstaedt, *Curious History of Relativity*, 247–48.

"on the fingers": Infeld, *Conférence internationale*, xv–xvi.

up-and-coming physicists: Ferreira, *Perfect Theory*, 81.

"Because there are no experiments": Feynmann, *Lectures on Gravitation*, xxvii.

"We who worked in this field were looked upon rather askance by other physicists": Infeld, *Conférence internationale*, xv.

Chapter 7: I Could Not Have Picked a More Exciting Time in Which to Become a Physicist

lay down an edict: DeWitt, "Quantum Gravity," 414.

"Within a few years": Hawking and Israel, *Three Hundred Years of Gravitation*, 250.

"Babson was fascinated": Kaiser, "Making Theory," 576. I am indebted to David Kaiser for making me aware of Roger Babson's unusual contribution to gravitational physics.

"through the Great Depression": Kaiser, "Making Theory," 577.

"oldest sister drowned": Kaiser, "Making Theory," 582–83. Babson later lost a grandson as well during a water rescue.

"It is to remind students": Kaiser, "Making Theory," 574.

Similar stones: See http://en.wikipedia.org/wiki/Gravity_Research_Foundation.

"[Tufts] legend": Kaiser, "Making Theory," 574.

"Mad men and quacks": Rickles, "Chapel Hill Conference," 11.

"waste of time": DeWitt, "New Directions for Research," 30.

Agnew Bahnson: Rickles, "Chapel Hill Conference," 9.

"no connection": Rickles, "Chapel Hill Conference," 13.

"landmark": Kaiser, "Making Theory," 592.

"By organizing conferences": Kaiser, "Making Theory," 594.

"set a record": Dyson, "John Archibald Wheeler," 126.

" 'one-legged men' ": APS, Wheeler Papers, box 18, Misner folder 1, Wheeler to Kenneth Case, 17 January 1964.

"There is no such thing": APS, Wheeler Papers, box 149, folder 12, Bekenstein to Wheeler, 23 August 1976.

"Before anyone else": Dyson, "John Archibald Wheeler," 128.

Wheeler grew up: Wheeler and Ford, *Geons, Black Holes,* 71–81.

"bent on making": Wheeler and Ford, *Geons, Black Holes,* 86.

"It is no exaggeration": Wheeler and Ford, *Geons, Black Holes,* 92–93.

"It was a non stop flight": APS, Wheeler Papers, box 133, Wheeler interview with Jeremy Bernstein, folder 1, p. 26.

"Vatican of physics": Wheeler and Ford, *Geons, Black Holes,* 142.

Publishing a paper with Bohr in 1939: Bohr and Wheeler, "Mechanism of Nuclear Fission."

"I finally had a calling": Wheeler and Ford, *Geons, Black Holes,* 159.

Just half an hour earlier: APS, Wheeler Papers, Relativity notebook, vol. 39. Before that, general relativity had been taught in the Princeton mathematics department. The Harvard physics department didn't add general relativity to its curriculum until 1967. See Kaiser, "A Ψ Is Just a Ψ?," 321–22.

"I wanted to teach relativity": Wheeler and Ford, *Geons, Black Holes,* 228.

"headed toward a complex thicket": APS, Wheeler Papers, box 184, Mentor and Sounding Board folder, p. 6.

pondering whether curved space: AIP, Kenneth Ford interview with John Wheeler, Section IX, 4 March 1994.

"I was looking": Wheeler and Ford, *Geons, Black Holes,* 229.

"He seemed to enjoy": Wheeler and Ford, *Geons, Black Holes,* 106–7.

"until it becomes a cinder": Wheeler and Ford, *Geons, Black Holes,* 229.

"no more than a superstition": Hoyle et al., "Relativistic Astrophysics," 909.

"free-wheeling talk sessions": Wheeler and Ford, *Geons, Black Holes,* 21.

book titled *Not Crazy Enough:* Conniff, "Johnny Wheeler's Space Odyssey," 14.

"He was so playful": Interview with Robert Fuller, 11 September 2013.

ultimately settling down: Thorne, *Black Holes and Time Warps,* 210.

"electromagnetic, gravitational, or neutrinos": Harrison, Wakano, and Wheeler, "Matter-Energy at High Density," 140.

"which lies at the untamed frontier": Harrison, Wakano, and Wheeler, "Matter-Energy at High Density," 137.

"Would not the simplest assumption": Harrison, Wakano, and Wheeler, "Matter-Energy at High Density," 147–48.

"It is very difficult": Harrison, Wakano, and Wheeler, "Matter-Energy at High Density," 148.

"green graduate student . . . I emerged": Klauder, *Magic Without Magic,* 231.

"In Western circles": Hawking and Israel, *Three Hundred Years of Gravitation,* 231.

"a body must tend": Hawking and Israel, *Three Hundred Years of Gravitation,* 231.

awarded a doctorate: Thorne, *Black Holes and Time Warps,* 222.

"the physics of stars": Sakharov, *Memoirs,* 102.

"intellectual progeny . . . Wheeler was a charismatic": Thorne, *Black Holes and Time Warps,* 261.

"Zel'dovich was the hard-driving player": Thorne, *Black Holes and Time Warps,* 261.

"When the mass": Thorne, *Black Holes and Time Warps,* 239.

"I had many discussions": APS, Wheeler Papers, box 133, transcript of Wheeler interview with Jeremy Bernstein, folder 1, p. 147.

the very same answer: Thorne, *Black Holes and Time Warps,* 240–41.

"frozen star": Thorne, *Black Holes and Time Warps,* 255.

"you cannot appreciate": Thorne, *Black Holes and Time Warps,* 244.

developed a new reference frame: see Finkelstein, "Past-Future Asymmetry."

he had joined: Hawking and Israel, *Three Hundred Years of Gravitation,* 238.

Kruskal arrived at: Kruskal, "Maximal Extension of Schwarzschild Metric."

Wheeler finally realized: Hawking and Israel, *Three Hundred Years of Gravitation,* 238.

both Finkelstein and Kruskal: Eisenstaedt, *Curious History of Relativity,* 293.

"The field of gravitation": Eisenstaedt, *Curious History of Relativity,* 299.

Beckedorff's treatise: Beckedorff, "Terminal Configurations of Stellar Evolution."

"Even if you sent": interview with Misner, 25 November 2013.

"But it's gone": interview with Misner.

"forever outside": Rindler, "Visual Horizons in World-Models," 663.

Evgeny Lifshitz and Isaak Khalatnikov: Lifshitz and Khalatnikov, "Singularities of Cosmological Solutions, I," and "Singularities of Cosmological Solutions, II."

"A black hole has no hair": Wheeler and Ford, *Geons, Black Holes,* 297.

"The matter of the core": Wheeler, "Our Universe," 5.

"When I got back to Cambridge": AIP, Alan Lightman interview with Roger Penrose, 24 January 1989.

"ridiculous and mysterious": said by Penrose during a roundtable discussion, "Recollections of the Relativistic Astrophysics Revolution," 27th Texas Symposium, Dallas, 11 December 2013.

published a theorem in *Physical Review Letters:* Penrose, "Gravitational Collapse."

"most influential development": Hawking and Israel, *Three Hundred Years of Gravitation,* 253.

"Deviations from spherical symmetry": Penrose, "Gravitational Collapse," 58.

"Cheshire cat": Wheeler, "Our Universe," 9.

"With this prediction": Wheeler, "Our Universe," 11.

"one final escape hatch": Thorne, *Black Holes and Time Warps,* 296–98.

"little relevance for the real Universe": Thorne, *Black Holes and Time Warps,* 268.

Chapter 8: It Was the Weirdest Spectrum I'd Ever Seen

Model-T Ford wheels: Sullivan, "Karl Jansky," 42.

"Jansky's merry-go-round": Kraus, "Karl Guthe Jansky's Serendipity," 58.

After a year of detective work: see Jansky, "Electrical Disturbances."

"star noise": Friis, "Karl Jansky," 842.

"result of some form of intelligence": "New Radio Waves Traced to Centre of the Milky Way."

"sounded like steam": "Radio Waves Heard from Remote Space."

"some sort of thermal agitation": Jansky, "Source of Interstellar Interference," 1162.

sent his results: Reber, "Cosmic Static" (1940).

first map: Reber, "Cosmic Static" (1944).

"The world of decibels": Sullivan, "Karl Jansky," 54.

"the most eventful era": Hawking and Israel, *Three Hundred Years of Gravitation,* 240.

ten million suns: Thorne, *Black Holes and Time Warps,* 339; see also Burbidge, "Possible Sources of Radio Emission," and "Theoretical Explanation of Radio Emission."

"Nuclear fuel's efficiency": Thorne, *Black Holes and Time Warps,* 340.

pinpoint of light: "First True Radio Star?," 148.

"I took a spectrum": Thorne, *Black Holes and Time Warps,* 335.

Astronomers couldn't even find: Hawking and Israel, *Three Hundred Years of Gravitation,* 243; also "First True Radio Star?," 148.

"ludicrous": Robinson, Schild, and Schücking, *Quasi-Stellar Sources,* xv.

sitting at his desk: Bartusiak, *Thursday's Universe,* 151.

rushing through space: Schmidt, "3C 273."

quasars: Physicist Hong-Yee Chiu coined the term, first using it in a 1964 *Physics Today* article he wrote on the First Texas Symposium. See Chiu, "Gravitational Collapse."

"*The Astrophysical Journal*": Schmidt, "Space Distribution," 371.

"The insult": Bartusiak, *Thursday's Universe,* 152.

old photographic plates: Hawking and Israel, *Three Hundred Years of Gravitation,* 246.

"The discovery of quasars": Schmidt, "Discovery of Quasars," 351.

"to the relativity limit": Hoyle and Fowler, "Nature of Strong Radio Sources," 535.

Vitaly L. Ginzburg: Ginzburg, "Nature of the Radio Galaxies."

Funders for this large gathering: Robinson, Schild, and Schücking, *Quasi-Stellar Sources,* iv.

antigravity devices: see Kaiser, "Making Theory," chap. 10, "Roger Babson and the Rediscovery of General Relativity."

"For more than ten years": Robinson, Schild, and Schücking, *Quasi-Stellar Sources,* v.

"I look forward": APS, Wheeler Papers, box 20, Penrose folder, Roger Penrose to Wheeler, 9 September 1963.

Chapter 9: Why Don't You Call It a Black Hole?

"for people who would recognize": Schücking, "First Texas Symposium," 48.

"Give a little spice to life": Schücking, "First Texas Symposium," 48.

"Nobody knows": Schücking, "First Texas Symposium," 49.

"Texanized": Schücking, "First Texas Symposium," 49.

"Particularly valuable": Schücking, "First Texas Symposium," 48–49.

"We fixed that": Schücking, "First Texas Symposium," 50.

"Relativity was the sleeping beauty": Renn, roundtable discussion, "Recollections of the Relativistic Astrophysics Revolution," 27th Texas Symposium, Dallas, 11 December 2013.

welcomed the conferees: Hawking and Israel, *Three Hundred Years of Gravitation*, 245.

synchronize our watches: Schücking, "First Texas Symposium," 50.

"Many of those": Green, "Dallas Conference," 80.

nine quasars: Green, "Dallas Conference," 84.

"magnificent cultural ornaments": Robinson, Schild, and Schücking, *Quasi-Stellar Sources*, 470.

"Among the problems": Robinson, Schild, and Schücking, *Quasi-Stellar Sources*, v.

Thousands of pockets: Green, "Dallas Conference," 82.

"We ignore": Robinson, Schild, and Schücking, *Quasi-Stellar Sources*, 17.

Hoyle and Fowler went on: Robinson, Schild, and Schücking, *Quasi-Stellar Sources*, 27.

That meant the quasar: Green, "Dallas Conference," 83.

a third of a light-year: Chiu, "Gravitational Collapse," 26.

only a few light-years wide: Green, "Dallas Conference," 83.

Freeman Dyson emphasized this point: Green, "Dallas Conference," 84.

Zel'dovich and Novikov . . . and Salpeter: see Zel'dovich and Novikov, "Gravitational Collapse," and Salpeter, "Accretion of Interstellar Matter."

gas clouds raining down: Robinson, Schild, and Schücking, *Quasi-Stellar Sources*, 437.

Kerr's presentation: Schücking, "First Texas Symposium," 50.

"in majestic isolation": Gamow, *Gravity*, 135.

"A new and very able": APS, Wheeler Papers, box 18, Misner folder 1, Wheeler to Kenneth Case, 17 January 1964.

Robert Pound and Glen Rebka finally measured: Pound and Rebka, "Apparent Weight of Photons."

"was so bad": Kerr, after-dinner talk, Kerr Conference, Potsdam, Germany, 4 July 2013; see http://www.kerr-conference.org/content/videoclip-archive.

his dissertation considered: Melia, *Cracking the Einstein Code*, 64.

"The next few weeks": Melia, *Cracking the Einstein Code,* 70.

"I wanted to find": Interview with Kerr, 12 December 2013.

"Alfred, it's spinning": Melia, *Cracking the Einstein Code,* 75; further description of this moment is also found in Kerr, "Discovering the Kerr and Kerr-Schild Metrics," 19.

"frame dragging": see Lense and Thirring, "On the Influence of the Proper Rotation."

"Cutting through": Melia, *Cracking the Einstein Code,* 75.

"I do not remember": Kerr, "Discovering the Kerr and Kerr-Schild Metrics," 21.

quickly offered: Melia, *Cracking the Einstein Code,* 76.

His final paper: Kerr, "Gravitational Field of a Spinning Mass."

bathe its campus tower: Kerr, Kerr Conference, Potsdam, Germany, 4 July 2013.

"lead balloon": Kerr, "The Kerr Solution at the First Texas Symposium 1963," 27th Texas Symposium, 11 December 2013.

"explain the large energies": Kerr, "Gravitational Field of a Spinning Mass," 99.

Achilles Papapetrou: Thorne, *Black Holes and Time Warps,* 342.

two surfaces defined: Interview with Kerr, 12 December 2013.

Penrose fully demonstrated: Penrose, "Gravitational Collapse."

Stephen Hawking, Brandon Carter, and David Robinson: see Hawking, "Black Holes in General Relativity"; Carter, "Axisymmetric Black Hole"; and Robinson, "Uniqueness of the Kerr Black Hole."

"the most shattering experience": Chandrasekhar, *Truth and Beauty,* 54.

Siraj's men incarcerated: Wolpert, *New History of India,* 179.

"Well, after I used that phrase": Bartusiak, *Einstein's Unfinished Symphony,* 62.

their existence remained a well-kept secret: Interview with Joseph Taylor, 11 December 2013.

The pulsar conference: e-mail communication with Stephen Maran, 27 May 2014.

Wheeler's name: see Brancazio and Cameron, *Supernovae and Their Remnants.*

What is indisputable: Wheeler, "Our Universe," 8–9.

"gravitational collapse": Rosenfeld, "What Are Quasi-Stellars?," 11.

Rosenfeld is sure: phone interview with Rosenfeld, 2012.

"space may be peppered": Ewing, "'Black Holes' in Space," 39.

"To the astonished audience": a letter dated 25 May 2009 describing Chiu's knowledge on the origin of the term *black hole* was sent by Chiu to *Physics Today.* It was not published, but Chiu kindly provided a copy.

His sons recall: an e-mail from John Dicke to Loyola University physicist Martin McHugh, with the kind permission of both to use it.

"He simply started": Thorne, *Black Holes and Time Warps,* 256.

an expression so exotic: For examples, see Kafka, "Possible Sources of Gravitational Radiation," 134, and Sullivan, "Pulsations from Space."

"He accused me": Wheeler and Ford, *Geons, Black Holes,* 297.

"Thus *black hole* seems the ideal name": Wheeler and Ford, *Geons, Black Holes,* 297.

"The advent of the term": Wheeler, *Journey into Gravity,* 211.

Chapter 10: Medieval Torture Rack

"There is a Chinese proverb": e-mail communication with Hong-Yee Chiu, 3 January 2014.

"a notable migration": Alexander, "Science Rediscovers Gravity," 101.

"Particle physics was in a mess": e-mail communication with Alan Lightman, 21 June 2014.

Black holes were proclaimed: Bartusiak, "Celestial Zoo," 108.

"Of all the concepts": Sullivan, "Probing the Mystery."

"black-hole disposal units": "Those Baffling Black Holes."

"black holes are out of sight": "Those Baffling Black Holes."

"noodlized": Wheeler, *Cosmic Catastrophes,* 182.

"mass without matter": Wheeler, "Lesson of the Black Hole," 25.

the mass of five billion suns: Wheeler, "Lesson of the Black Hole," 33.

"On first approach": Wheeler, "Lesson of the Black Hole," 25.

"As long as this viewpoint prevailed": Price and Thorne, "Introduction: The Membrane Paradigm," 2.

Chapter 11: Whereas Stephen Hawking Has Such a Large Investment in General Relativity and Black Holes and Desires an Insurance Policy

"There is about as little hope": Wheeler, "Superdense Star," 195.

If that companion emits no light: Zel'dovich and Guseynov, "Collapsed Stars in Binaries."

"not because they 'wish' ": Zel'dovich and Guseynov, "Collapsed Stars in Binaries," 840.

Working with astronomer Virginia Trimble, Kip Thorne: Thorne *Black Holes and Time Warps,* 306–7.

announce itself by the glow: Zel'dovich and Novikov, "Gravitational Collapse."

"[This] proposed method of searching": Zel'dovich and Novikov, "Gravitational Collapse."

light of a full Moon: Tucker and Giacconi, *X-Ray Universe,* 42.

three large Geiger counters: Giacconi et al., "Evidence for X Rays," 439.

"trying to get support": AIP, Richard Hirsh interview with Riccardo Giacconi, 12 July 1976.

worried their instrument: Tucker and Giacconi, *X-Ray Universe,* 43.

"The discovery of pulsars": Thorne, "Nonspherical Gravitational Collapse," 191.

"X-Ray Scanning Satellite": Sullivan, "X-Ray Scanning Satellite."

radio and optical astronomers: see Wade and Hjellming, "Position and Identification," and Bolton, "Identification of Cygnus X-1."

orbital measurements: Bolton, "Dimensions of the Binary System."

"Whereas Stephen Hawking": Hawking and Israel, *Three Hundred Years of Gravitation,* 249.

"Late one night": Thorne, *Black Holes and Time Warps,* 315.

the higher the bulge mass: Irion, "Quasar in Every Galaxy?" 42.

form a doughnutlike ring: Lynden-Bell, "Galactic Nuclei."

"The gas gauge says 'empty' ": Bartusiak, *Thursday's Universe,* 163.

Chapter 12: Black Holes Ain't So Black

Hawking himself: Ferguson, *Stephen Hawking,* 30.

"One evening": Hawking, *Brief History of Time,* 99.

"There was no precise definition": Hawking, *Brief History of Time,* 99.

The session devoted: Hajicek, "Fifth Texas Symposium," 178.

entropy was *zero:* Hawking and Israel, *Three Hundred Years of Gravitation,* 262.

"I drew comfort": Bekenstein, "Black Hole Thermodynamics," 28.

"I suspect": APS, Wheeler Papers, box 4, Bekenstein folder, Bekenstein to Wheeler, 25 September 1973.

"Such an identification": Bekenstein, "Black Holes and Entropy," 2338.

"unambiguously zero": Hawking and Israel, *Three Hundred Years of Gravitation,* 262.

"I was motivated": Hawking, *Brief History of Time,* 104.

all black holes—spinning or not: Hawking, *Brief History of Time,* 104–5.

"Black holes ain't so black": Hawking, *Brief History of Time,* 99.

"Black Hole Explosions?": Hawking, "Black Hole Explosions?"

one million megaton hydrogen bombs: Hawking, "Black Hole Explosions?" 30.

"aroused strong opposition": Hawking and Israel, *Three Hundred Years of Gravitation*, 265.

"Sorry, Stephen": Boslough, *Stephen Hawking's Universe*, 70.

"I was probably the most pleased": Bekenstein, "Black Hole Thermodynamics," 28.

less than a millionth of a degree: Wheeler, *Journey into Gravity*, 222.

Some recent models even suggest: see Marolf and Polchinski, "Gauge-Gravity Duality."

Wheeler held out the hope: Wheeler and Ford, *Geons, Black Holes*, 229.

"the core of a black hole": Wheeler and Ford, *Geons, Black Holes*, 295.

Chapter 13: Epilogue

LIGO: the history of gravitational-wave astronomy and the development of LIGO can be found in Bartusiak, *Einstein's Unfinished Symphony*.

Einstein first wrote: Einstein, "Näherungsweise Integration," and "Über Gravitationswellen."

The rate of their orbital decay: a full discussion of the Hulse-Taylor binary can be found in Taylor, "Binary Pulsars."

"We didn't know that neutron stars": Joseph Taylor, Plenary Presentation I, 27th Texas Symposium, Dallas, 9 December 2013.

"All the science fiction I need is right out there in front of us": Conniff, "Johnny Wheeler's Space Odyssey."

Bibliography

Adams, W. S. "An A-Type Star of Very Low Luminosity." *Publications of the Astronomical Society of the Pacific* 26 (1914): 198.

———. "The Spectrum of the Companion of Sirius." *Publications of the Astronomical Society of the Pacific* 27 (1915): 236–37.

Alexander, Tom. "Science Rediscovers Gravity." *Fortune* (December 1969): 100–104, 187–88.

Anderson, A. "On the Advance of the Perihelion of a Planet, and the Path of a Ray of Light in the Gravitation Field of the Sun." *Philosophical Magazine* 39 (1920): 626–28.

Baade, W., and F. Zwicky. "Cosmic Rays from Super-Novae." *Proceedings of the National Academy of Sciences* 20 (May 1934): 259–63.

———. "On Super-Novae." *Proceedings of the National Academy of Sciences* 20 (May 1934): 254–59.

Bartusiak, Marcia. "A Beast in the Core." *Astronomy* (July 1998): 42–47.

———. "Celestial Zoo." *Omni* (December 1982): 106–13.

———. *Einstein's Unfinished Symphony.* Washington, DC: Joseph Henry, 2000.

———. *Through a Universe Darkly.* New York: HarperCollins, 1993.

———. *Thursday's Universe.* New York: Times Books, 1986.

Beckedorff, D. L. "Terminal Configurations of Stellar Evolution." AB thesis, Princeton University, Department of Mathematics, 1962.

Begelman, Mitchell, and Martin Rees. *Gravity's Fatal Attraction: Black Holes in the Universe.* New York: Scientific American Library, 1996.

Bekenstein, Jacob D. "Black Holes and Entropy." *Physics Review D* 7 (1973): 2333–46.

————. "Black Hole Thermodynamics." *Physics Today* 33 (January 1980): 24–31.

Bessel, F. W. "On the Variations of Proper Motions of Procyon and Sirius." *Monthly Notices of the Royal Astronomical Society* 6 (1844): 136–41.

Beyerchen, Alan. *Scientists Under Hitler: Politics and the Physics Community in the Third Reich.* New Haven: Yale University Press, 1977.

Bohr, Niels, and John Archibald Wheeler. "The Mechanism of Nuclear Fission." *Physical Review* 56 (1939): 426–50.

Bolton, C. T. "Dimensions of the Binary System HDE 226868 = Cygnus X-1." *Nature Physical Science* (11 December 1972): 124–27.

————. "Identification of Cygnus X-1 with HDE 226868." *Nature* 235 (1972): 271–73.

Bond, George P. "On the Companion of Sirius." *American Journal of Science* 33 (1862): 286–87.

Born, Max. "Physics and Relativity." In *Fünfzig Jahre Relativitätstheorie,* ed. André Mercier and Michel Kervaire, 244–60. *Helvetia Physica Acta,* Supplement 4. Basel: Birkhäuser, 1956.

Boslough, John. *Stephen Hawking's Universe.* New York: W. Morrow, 1985.

Braccesi, Alessandro. "Revisiting Fritz Zwicky." In *Modern Cosmology in Retrospect,* ed. B. Bertotti, R. Balbinot, S. Bergia, and A. Messina, 415–23. Cambridge: Cambridge University Press, 1990.

Brancazio, Peter J., and A. G. W. Cameron, eds. *Supernovae and Their Remnants: Proceedings of the Conference on Supernovae, Held at the Goddard Institute for Space Studies, NASA 1967.* New York: Gordon and Breach Science, 1969.

Brewster, David. *Memoirs of the Life, Writings and Discoveries of Sir Isaac Newton,* vol. 1. Edinburgh: Thomas Constable, 1855.

Burbidge, G. R. "Possible Sources of Radio Emission in Clusters of Galaxies." *Astrophysical Journal* 28 (July 1958): 1–8.

————. "The Theoretical Explanation of Radio Emission." In *Radio Symposium on Radio Astronomy,* ed. Ronald N. Bracewell, 541–53. Stanford, CA: Stanford University Press, 1959.

Carter, B. "Axisymmetric Black Hole Has Only Two Degrees of Freedom." *Physical Review Letters* 26 (1971): 331–33.

Cassidy, David C. *J. Robert Oppenheimer and the American Century.* Baltimore: Johns Hopkins University Press, 2005.

Chadwick, J. "Possible Existence of a Neutron." *Nature* 129 (1932): 312.

Chandrasekhar, S. "Beauty and the Quest for Beauty in Science." *Physics Today* 63 (December 2010): 57–62.

———. "The Black Hole in Astrophysics: The Origin of the Concept and Its Role." *Contemporary Physics* 15 (1974): 1–24.

———. "The Density of White Dwarfs." *Philosophical Magazine* 11 (1931): 592–96.

———. "The Highly Collapsed Configurations of a Stellar Mass." *Monthly Notices of the Royal Astronomical Society* 91 (1931): 456–66.

———. "The Highly Collapsed Configurations of a Stellar Mass (Second Paper)." *Monthly Notices of the Royal Astronomical Society* 95 (1935): 207–25.

———. "The Maximum Mass of Ideal White Dwarfs." *Astrophysical Journal* 74 (1931): 81–82.

———. "Some Remarks on the State of Matter in the Interior of Stars." *Zeitschrift für Astrophysik* 5 (1932): 321–27.

———. "Stellar Configurations with Degenerate Cores." *Observatory* 57 (1934): 373–77.

———. *Truth and Beauty: Aesthetics and Motivations in Science.* Chicago: University of Chicago Press, 1987.

———. "Verifying the Theory of Relativity." *Notes and Records of the Royal Society* 30 (January 1976): 249–60.

———. "The White Dwarfs and Their Importance for Theories of Stellar Evolution." In *Novae and White Dwarfs*, vol. 3, ed. Knut Lundmark et al., 239–48. Colloque International d'Astrophysique, 17–23 July 1939, Paris. Paris: Hermann, 1941.

Chiu, Hong-Yee. "Gravitational Collapse." *Physics Today* 17 (May 1964): 21–34.

Ciufolini, Ignazio, and Richard A. Matzner, eds. *General Relativity and John Archibald Wheeler.* Dordrecht: Springer, 2010.

Cohen, I. B. "Newton." *Dictionary of Scientific Biography,* vol. 10. New York: Scribner's, 1974.

Comins, Neil F., and William J. Kaufmann. *Discovering the Universe: From the Stars to the Planets.* New York: Macmillan, 2008.

Conniff, James C. G. "Johnny Wheeler's Space Odyssey." *Today—The Philadelphia Inquirer,* 16 March 1975.

Crossley, Richard. "Mystery at the Rectory: Some Light on John Michell." *Annual Report, 2003,* Yorkshire Philosophical Society.

DeVorkin, David. H. "Steps Toward the Hertzsprung-Russell Diagram." *Physics Today* 31 (March 1978): 32–39.

DeWitt, Bryce. "New Directions for Research in the Theory of Gravitation." Winning paper submitted to Gravity Research Foundation's essay contest in 1953. See http://www.gravityresearchfoundation.org/pdf/awarded/1953/dewitt.pdf.

———. "Quantum Gravity: Yesterday and Today." *General Relativity and Gravitation* 41 (2009): 413–19.

"Discussion of Papers by A. S. Eddington and E. A. Milne." *Observatory* 58 (1935): 37–39.

Dyson, Freeman. "Chandrasekhar's Role in 20th-Century Science." *Physics Today* 63 (December 2010): 44–48.

———. "John Archibald Wheeler." *Proceedings of the American Philosophical Society* 154 (March 2010): 126–29.

Earman, John, and Jean Eisenstaedt. "Einstein and Singularities." *Studies in the History and Philosophy of Modern Physics* 30 (1999): 185–235.

Eddington, Arthur. *The Internal Constitution of the Stars.* Cambridge: Cambridge University Press, 1926.

———. "On 'Relativistic Degeneracy.'" *Monthly Notices of the Royal Astronomical Society* 95 (1935): 194–206.

———. "On the Relation Between the Masses and the Luminosities of Stars." *Observatory* 47 (1924): 107–14.

———. *Space, Time, and Gravitation.* Cambridge: Cambridge University Press, 1920.

———. *Stars and Atoms.* Oxford: Clarendon Press, 1927.

Einstein, Albert. "Autobiographical Notes." In *Albert Einstein: Philosopher-Scientist,* ed. Paul Arthur Schilpp, 1–95. Evanston, IL: Library of Living Philosophers, 1949.

———. *The Berlin Years, 1914–1917.* Volume 6 of *The Collected Papers of Albert Einstein,* trans. Alfred Engel. Princeton, NJ: Princeton University Press, 1997.

———. *The Berlin Years, Correspondence, 1914–1918.* Volume 8 of *The Collected Papers of Albert Einstein,* trans. Ann M. Hentschel. Princeton, NJ: Princeton University Press, 1998.

———. *The Berlin Years, Correspondence, May–December 1920, and Supplementary Correspondence, 1909–1920.* Volume 10 of *The Collected Papers of Albert Einstein,* ed. Diana Kormos Buchwald et al. Princeton, NJ: Princeton University Press, 2006.

———. *The Early Years, 1879–1902.* Volume 1 of *The Collected Papers of Albert Einstein,* trans. Anna Beck. Princeton, NJ: Princeton University Press, 1987.

———. *The Meaning of Relativity,* 3rd ed. Princeton, NJ: Princeton University Press, 1950.

———. "Näherungsweise Integration der Feldgleichungen der Gravitation." *Sitzungsberichte der Königlich Preussischen Akademie der Wissenschaften* (1916): 688–96.

———. "On a Stationary System with Spherical Symmetry Consisting of Many Gravitating Masses." *Annals of Mathematics* 40 (1939): 922–36.

———. "On the Influence of Gravity on the Propagation of Light." *Annalen der Physik* 35 (1911): 898–908.

———. *The Swiss Years, Correspondence, 1902–1914.* Volume 5 of *The Collected Papers of Albert Einstein,* trans. Anna Beck. Princeton, NJ: Princeton University Press, 1995.

———. "Über Gravitationswellen." *Sitzungsberichte der Königlich Preussischen Akademie der Wissenschaften* (1918): 154–67.

———. "Zur Elektrodynamik beweger Körper." *Annalen der Physik* 17 (1905): 891–921.

Eisenstaedt, Jean. *The Curious History of Relativity.* Princeton, NJ: Princeton University Press, 2006.

———. "Light and Relativity, a Previously Unknown Eighteenth-Century Manuscript by Robert Blair (1748–1828)." *Annals of Science* 62 (2005): 347–76.

Ewing, Ann. " 'Black Holes' in Space." *Science News Letter,* 18 January 1964, 39.

"Fascination with Celestial Events Is Deeply Ingrained." *Japan Report,* vols. 21–22. Japan Information Center, Consulate General of Japan, 1975.

Ferguson, Kitty. *Stephen Hawking: An Unfettered Mind.* New York: Palgrave Macmillan, 2012.

Ferrari, Valeria. In "Some Memories of Chandra." *Physics Today* 63 (December 2010): 49–53.

Ferreira, Pedro G. *The Perfect Theory.* Boston: Houghton Mifflin Harcourt, 2014.

Feynman, R. P., et al. *Feynman Lectures on Gravitation*. Reading, MA: Addison-Wesley, 1995.

Finkelstein, David. "Past-Future Asymmetry of the Gravitational Field of a Point Particle." *Physical Review* 110 (1958): 965–67.

"First True Radio Star?" *Sky and Telescope* 21 (March 1961): 148.

Fölsing, Albrecht. *Albert Einstein: A Biography*. New York: Viking, 1997.

Fowler, Ralph H. "On Dense Matter." *Monthly Notices of the Royal Astronomical Society* 87 (December 1926): 114–22.

———. *Statistical Mechanics*. Cambridge: Cambridge University Press, 1929.

Friedman, John. In "Some Memories of Chandra." *Physics Today* 63 (December 2010): 49–53.

Friis, Harold. "Karl Jansky: His Career at Bell Telephone Laboratories." *Science* (1965): 841–42.

Gamow, George. *Gravity*. Mineola, NY: Dover, 2002.

———. *Structure of Atomic Nuclei and Nuclear Transformations*. Oxford: Clarendon Press, 1937.

Giacconi, Riccardo, et al. "Evidence for X Rays from Sources Outside the Solar System." *Physical Review Letters* 9 (1962): 439–43.

Gillispie, Charles Coulston. *Pierre-Simon Laplace, 1749–1827: A Life in Exact Science*. Princeton, NJ: Princeton University Press, 1997.

Ginzburg, V. L. "The Nature of the Radio Galaxies." *Soviet Astronomy* 5 (1961): 282–83.

Gleiser, Marcelo. "Relativity's Later Years." *Journal for the History of Astronomy* 38 (November 2007): 522–24.

Green, Louis C. "Dallas Conference on Super Radio Sources." *Sky and Telescope* 27 (February 1964): 80–84.

Hajicek, P. "Report on the Fifth Texas Symposium on Relativistic Astrophysics." *General Relativity and Gravitation* 2 (1971): 173–81.

Halley, Edmund [or Edmond]. *A Synopsis of the Astronomy of Comets*. London: John Senex, 1705.

Halpern, Paul, and Paul Wesson. *Brave New Universe: Illuminating the Darkest Secrets of the Universe*. Washington, DC: Joseph Henry Press, 2006.

Hardin, Clyde. "The Scientific Work of the Reverend John Michell." *Annals of Science* 22 (1966): 27–47.

Harrison, B. K., M. Wakano, and J. A. Wheeler. "Matter-Energy at High Density; End Point of Thermonuclear Evolution." In *La Structure et l'évolution de l'univers*. Onzième Conseil de Physique Solvay. Brussels: Stoops, 1958.

Harrison, B. Kent, et al. *Gravitation Theory and Gravitational Collapse*. Chicago: University of Chicago Press, 1965.

Hawking, S. W. "Black Hole Explosions?" *Nature* 248 (1974): 30–31.

———. "Black Holes in General Relativity." *Communications in Mathematical Physics* 25 (1972): 152–66.

———. *A Brief History of Time: From the Big Bang to Black Holes*. New York: Bantam Books, 1988.

Hawking, Stephen, and Werner Israel, eds. *Three Hundred Years of Gravitation*. Cambridge: Cambridge University Press, 1989.

Herschel, William. "Catalogue of Double Stars." *Philosophical Transactions of the Royal Society of London* 75 (1785): 40–126.

Hoffmann, Banesh. *Albert Einstein: Creator and Rebel*. New York: Viking, 1972.

Holberg, J. B., and F. Wesemael. "The Discovery of the Companion of Sirius and Its Aftermath." *Journal of the History of Astronomy* 38 (2007): 167.

Hoyle, F., and William A. Fowler. "Nature of Strong Radio Sources." *Nature* 197 (1963): 533–35.

Hoyle, F., et al. "On Relativistic Astrophysics." *Astrophysical Journal* 139 (1964): 909–28.

Hubble, Edwin. "A Spiral Nebula as a Stellar System, Messier 31." *Astrophysical Journal* 69 (1929): 103–58.

Hufbauer, Karl. "J. Robert Oppenheimer's Path to Black Holes." In *Reappraising Oppenheimer: Centennial Studies and Reflections,* ed. Cathryn Carson and David A. Hollinger, 31–47. Berkeley: Office for History of Science and Technology, University of California, Berkeley, 2005.

———. "Landau's Youthful Sallies into Stellar Theory: Their Origins, Claims, and Receptions." *Historical Studies in the Physical and Biological Sciences* 37 (2007): 337–354.

———. "Stellar Structure and Evolution, 1924–1939." *Journal for the History of Astronomy* 37 (2006): 203–27.

Infeld, L., ed. *Conférence internationale sur les théories relativistes de la gravitation,* Warsaw and Jablonna, 25–31 July 1962. Paris: Gauthier-Villars, 1964.

International Astronomical Union Circular No. 2826, 2 September 1975.

Irion, Robert. "A Quasar in Every Galaxy?" *Sky and Telescope* 112 (July 2006): 40–46.

Israel, Werner. "From White Dwarfs to Black Holes: The History of a Revolutionary Idea." *Queen's Quarterly* 95 (1988): 78–89.

———. "Imploding Stars, Shifting Continents, and the Inconstancy of Matter." *Foundations of Physics* 26 (1996): 595–616.

Jansky, Karl. "Electrical Disturbances Apparently of Extraterrestrial Origin." *Proceedings of the Institute of Radio Engineers* 21 (1933): 1387–98.

———. "A Note on the Source of Interstellar Interference." *Proceedings of the Institute of Radio Engineers* 23 (1935): 1162.

Jeffreys, H. "The Compressibility of Dwarf Stars and Planets." *Monthly Notices of the Royal Astronomical Society* 78 (1918): 183–84.

Jungnickel, Christa, and Russell McCormmach. *Cavendish: The Experimental Life.* Cranbury, NJ: Bucknell University Press, 1999.

Kafka, P. "Discussion of Possible Sources of Gravitational Radiation." *Mitteilungen der Astronomischen Gesellschaft* 27 (1969): 134–38.

Kaiser, David. "Making Theory: I. Producing Physics and Physicists in Postwar America." PhD diss., Harvard University, 2000.

———. "A Ψ Is Just a Ψ? Pedagogy, Practice, and the Reconstitution of General Relativity, 1942–1975." *Studies in History and Philosophy of Modern Physics* 29 (1998): 321–38.

Kelvin, Lord. *Baltimore Lectures: On Molecular Dynamics and the Wave Theory of Light.* London: C. J. Clay and Sons, 1904.

Kennefick, Daniel. *Traveling at the Speed of Thought.* Princeton, NJ: Princeton University Press, 2007.

Kerr, Roy Patrick. "Discovering the Kerr and Kerr-Schild Metrics." arXiv.org, arXiv:0706.1109v1 [gr-qc], General Relativity and Quantum Cosmology, 8 June 2007.

———. "Gravitational Collapse and Rotation." In *Quasi-Stellar Sources and Gravitational Collapse: Including the Proceedings of the First Texas Symposium on Relativistic Astrophysics,* ed. Ivor Robinson, Alfred Schild, and E. L. Schücking, 99–102. Chicago: University of Chicago Press, 1965.

———. "Gravitational Field of a Spinning Mass as an Example of Algebraically Special Metrics." *Physics Review Letters* 11 (September 1963): 237–38.

Klauder, John R., ed. *Magic Without Magic: John Archibald Wheeler, a Collection of Essays in Honor of His Sixtieth Birthday.* San Francisco: W. H. Freeman, 1972.

Kraus, John. "Karl Guthe Jansky's Serendipity, Its Impact on Astronomy and Its Lessons for the Future." *Serendipitous Discoveries in Radio Astronomy: Proceedings of a Workshop Held at the National Radio Astronomy Observatory, Green Bank, West Virginia on May 4, 5, 6, 1983,* ed. K. Kellermann and B. Sheets, 57–70. Green Bank, WV: National Radio Astronomy Observatory, 1983.

Kruskal, M. D. "Maximal Extension of Schwarzschild Metric." *Physical Review* 119 (1960): 1743–45.

Landau, L. "On the Theory of Stars." *Physikalische Zeitschrift der Sowjetunion* 1 (1932): 285–88.

———. "Origin of Stellar Energy." *Nature* 141 (1938): 333–34.

Laplace, P. S. *System of the World,* vol. 2. Trans. J. Pond. London: W. Flint, 1809.

Lense, J., and H. Thirring. "On the Influence of the Proper Rotation of Central Bodies on the Motions of Planets and Moons According to Einstein's Theory of Gravitation." *Physikalische Zeitschrift* 19 (1918): 156–63.

Lifshitz, E. M., and I. M. Khalatnikov. "On the Singularities of Cosmological Solutions of the Gravitational Equations, I." *Soviet Physics—Journal of Experimental and Theoretical Physics* 12, no. 1 (1961): 108, 558.

———. "On the Singularities of Cosmological Solutions of the Gravitational Equations, II." *Soviet Physics—Journal of Experimental and Theoretical Physics* 12, no. 3 (1961): 558.

Lodge, Sir Oliver. "On the Supposed Weight and Ultimate Fate of Radiation." *Philosophical Magazine* 41 (1921): 549–57.

Lorentz, H. A., et al. *The Principle of Relativity.* New York: Methuen, 1923.

Lundmark, K. "The Pre-Tychonic Novae." *Lund Observatory Circular* 8 (1932): 216–18.

Lundmark, Knut, et al., eds. *Novae and White Dwarfs.* Colloque International d'Astrophysique, 17–23 July 1939, Paris. Paris: Hermann, 1941.

Lynden-Bell, D. "Galactic Nuclei as Collapsed Old Quasars." *Nature* (1969): 690–94.

Marolf, Donald, and Joseph Polchinski. "Gauge-Gravity Duality and the Black Hole Interior." *Physical Review Letters* 111 (2013): 171301-1-5.

Matthews, Thomas A., and Allan R. Sandage. "Optical Identification of 3C 48, 3C 196, and 3C 286 with Stellar Objects." *Astrophysical Journal* 138 (1963): 30–56.

Maxwell, James Clerk. "A Dynamical Theory of the Electromagnetic Field." *Philosophical Transactions of the Royal Society of London* 155 (1865): 459–512.

———. "Introductory Lecture on Experimental Physics." In *The Scientific Papers of James Clerk Maxwell,* vol. 2, ed. W. D. Niven, 241–55. Cambridge: Cambridge University Press, 1890.

McClintock, Jeffrey. "Do Black Holes Exist?" *Sky and Telescope* (January 1988): 28–33.

McCormmach, Russell. "John Michell and Henry Cavendish: Weighing the Stars." *British Journal for the History of Science* 4 (December 1968): 126–55.

Melia, Fulvio. *Cracking the Einstein Code: Relativity and the Birth of Black Hole Physics.* Chicago: University of Chicago Press, 2009.

Michell, John. "An Inquiry into the Probable Parallax, and Magnitude of the Fixed Stars, from the Quantity of Light Which They Afford us, and the Particular Circumstances of Their Situation." *Philosophical Transactions of the Royal Society of London* 57 (1767): 234–64.

———. "On the Means of discovering the Distance, Magnitude, &c. of the Fixed Stars, in consequence of the Diminution of the Velocity of their Light, in case such a Diminution should be found to take place in any of them, and such other Data should be procured from Observations, as would be farther necessary for that Purpose." *Philosophical Transactions of the Royal Society of London* 74 (1784): 35–57.

Miller, Arthur I. *Empire of the Stars: Obsession, Friendship, and Betrayal in the Quest for Black Holes.* Boston: Houghton Mifflin, 2005.

Minkowski, H. "Space and Time." In H. A. Lorentz et al., *The Principle of Relativity,* 75–91. London: Methuen, 1923.

Misner, Charles. "Infinite Red-Shifts in General Relativity." In *The Nature of Time,* ed. T. Gold, 75–89. Ithaca, NY: Cornell University Press, 1967.

Mobberley, Martin. *Cataclysmic Cosmic Events and How to Observe Them.* New York: Springer, 2008.

Montgomery, Colin, Wayne Orchiston, and Ian Whittingham. "Michell, Laplace and the Origin of the Black Hole Concept." *Journal of Astronomical History and Heritage* 12, no. 2 (2009): 90–96.

Nauenberg, Michall. "Edmund C. Stoner and the Discovery of the Maximum Mass of White Dwarfs." *Journal for the History of Astronomy* 39 (2008): 297–312.

"New Radio Waves Traced to Centre of the Milky Way." *New York Times,* 5 May 1933.

Newton, Isaac. *The Principia.* Trans. I. Bernard Cohen and Anne Whitman. Berkeley: University of California Press, 1999.

Öpik, E. "On the Densities of Visual Binary Stars." *Astrophysical Journal* 44 (1916): 292–302.

Oppenheimer, J. R., and Robert Serber. "On the Stability of Stellar Neutron Cores." *Physical Review* 54 (1938): 540.

Oppenheimer, J. R., and H. Snyder. "On Continued Gravitational Contraction." *Physical Review* 56 (1939): 455–59.

Oppenheimer, J. R., and G. M. Volkoff. "On Massive Neutron Cores." *Physical Review* 55 (15 February 1939): 374–81.

Osterbrock, Donald E. *Walter Baade: A Life in Astrophysics.* Princeton, NJ: Princeton University Press, 2001.

Pais, Abraham. *"Subtle Is the Lord . . .": The Science and the Life of Albert Einstein.* Oxford: Oxford University Press, 1982.

Penrose, R. "Gravitational Collapse: The Role of General Relativity." *Rivista del Nuovo Cimento, Numero Speziale 1* (1969): 252–76.

———. "Gravitational Collapse and Spacetime Singularities." *Physical Review Letters* 14 (1965): 57–59.

Philip, A. G. Davis, and D. H. DeVorkin, eds. "In Memory of Henry Norris Russell." *Dudley Observatory Report* 13 (1977).

Piaggio, H. T. H., and J. Critchlow. "A Supposed Relativity Method of Determining the Size of a Gravitating Particle." *Philosophical Magazine,* 7th ser., 1 (1926): 67–71.

"Placing Chandra's Work in Historical Context." *Physics Today* 64 (July 2011): 8–10.

Pound, R. V., and G. A. Rebka Jr. "Apparent Weight of Photon." *Physical Review Letters* 4 (1960): 337–41.

"Radio Waves Heard from Remote Space." *New York Times*, 16 May 1933.

Reber, Grote. "Cosmic Static." *Astrophysical Journal* 91 (1940): 621–24.

———. "Cosmic Static." *Astrophysical Journal* 100 (1944): 279–87.

Rees, Martin, Remo Ruffini, and John Archibald Wheeler. *Black Holes, Gravitational Waves and Cosmology: An Introduction to Current Research.* New York: Gordon and Breach, 1974.

Rickles, Dean. "The Chapel Hill Conference in Context." In *The Role of Gravitation in Physics: Report from the 1957 Chapel Hill Conference,* ed. Dean Rickles and Cécile M. DeWitt. Edition Open Access; Max Planck Research Library for the History and Development of Knowledge, 2011. http://www.edition-open-access.de/sources/5/index.html.

Rindler, W. "Visual Horizons in World-Models." *Monthly Notices of the Royal Astronomical Society* 116 (1956): 662–77.

Robinson, D. C. "Uniqueness of the Kerr Black Hole." *Physical Review Letters* 34 (1975): 905–6.

Robinson, Ivor, Alfred Schild, and E. L. Schücking, eds. *Quasi-Stellar Sources and Gravitational Collapse: Including the Proceedings of the First Texas Symposium on Relativistic Astrophysics.* Chicago: University of Chicago Press, 1965.

Rosenfeld, Albert. "What Are Quasi-Stellars? Heavens' New Enigma." *Life,* 24 January 1964, 11–12.

Ruffini, Remo, and John A. Wheeler. "Introducing the Black Hole." *Physics Today* 24 (January 1971): 30–41.

Russell, Henry Norris. "Address by Professor Henry Norris Russel [sic]." In *Novae and White Dwarfs,* vol. 1, ed. Knut Lundmark et al., 1–5. Colloque International d'Astrophysique, 17–23 July 1939, Paris. Paris: Hermann, 1941.

Sakharov, Andrei. *Memoirs.* New York: Knopf, 1990.

Salpeter, E. E. "Accretion of Interstellar Matter by Massive Objects." *Astrophysical Journal* 140 (1964): 796–800.

Sampson, R. A. "On the Validity of the Principles of Relativity and Equivalence." *Monthly Notices of the Royal Astronomical Society* 80 (1919): 154–57.

Schaffer, Simon. "John Michell and Black Holes." *Journal for the History of Astronomy* 10 (1979): 42–43.

Schemmel, Matthias. "An Astronomical Road to General Relativity: The Continuity Between Classical and Relativistic Cosmology in the Work of Karl Schwarzschild." *Science in Context* 18 (2005): 451–78.

Schmidt, Maarten. "The Discovery of Quasars." In *Modern Cosmology in Retrospect*, ed. B. Bertotti et al., 347–54. Cambridge: Cambridge University Press, 1990.

———. "Space Distribution and Luminosity Functions of Quasars." *Astrophysical Journal* 162 (1970): 371–79.

———. "3C 273: A Star-Like Object with Large Red-Shift." *Nature* 197 (1963): 1040.

Schücking, Engelbert L. "The First Texas Symposium on Relativistic Astrophysics." *Physics Today* 42 (August 1989): 46–52.

Schwarzschild, K. "On the Gravitational Field of a Mass Point According to Einstein's Theory." *Sitzungsberichte der Königlich Preussischen Akademie der Wissenschaften zu Berlin, Phys.-Math. Klasse* (1916): 189–96.

———. "On the Gravitational Field of a Mass Point According to Einstein's Theory." Translated in *General Relativity and Gravitation* 35 (May 2003): 951–59.

———. "On the Gravitational Field of a Sphere of Incompressible Fluid According to Einstein's Theory." *Sitzungsberichte der Königlich Preussischen Akademie der Wissenschaften zu Berlin* (1916): 424–34.

———. "Ueber das zulässige Krümmungsmaass des Raumes." *Vierteljahrsschrift der Astronomischen Gesellschaft* 35 (1900): 337–47.

Shallit, Jeffrey. "Science, Pseudoscience, and the Three Stages of Truth." Unpublished paper, 2005. https://cs.uwaterloo.ca/~shallit/Papers/stages.pdf.

Shipman, Harry L. *The Restless Universe.* New York: Houghton Mifflin, 1978.

Smith, Alice Kimball, and Charles Weiner, eds. *Robert Oppenheimer: Letters and Recollections.* Cambridge, MA: Harvard University Press, 1980.

Stachel, John. *Einstein from "B" to "Z."* Boston: Birkhäuser, 2002.

Sullivan, A. M. "Music of the Spheres." *New York Times*, 26 August 1967.

Sullivan, Walter. *Black Holes: The Edge of Space, the End of Time.* Garden City, NY: Anchor, 1979.

———. "Probing the Mystery of the 'Black Holes.'" *New York Times*, 4 April 1971.

———. "Pulsations from Space." *New York Times*, 14 April 1968.

———. "An X-Ray Scanning Satellite May Have Discovered a 'Black Hole' in Space." *New York Times*, 1 April 1971.

Sullivan, Woodruff T., III. "Karl Jansky and the Beginnings of Radio Astron-
omy." In *Serendipitous Discoveries in Radio Astronomy: Proceedings of a
Workshop Held at the National Radio Astronomy Observatory, Green Bank,
West Virginia on May 4, 5, 6, 1983,* ed. K. Kellermann and B. Sheets, 39–56.
Green Bank, WV: National Radio Astronomy Observatory, 1983.

Taylor, Edwin F., and John Archibald Wheeler. *Spacetime Physics: Introduction to Special Relativity,* 2nd ed. New York: Macmillan, 1992.

Taylor, Joseph H., Jr. "Binary Pulsars and Relativistic Gravity." *Reviews of
Modern Physics* 66 (1994): 711–19.

Thorne, Kip S. *Black Holes and Time Warps: Einstein's Outrageous Legacy.*
New York: W. W. Norton, 1994.

———. "Nonspherical Gravitational Collapse: Does It Produce Black
Holes?" *Comments on Astrophysics and Space Physics* 2 (1970): 191–96.

Thorne, Kip S., Richard H. Price, and Douglas A. MacDonald, eds. *Black
Holes: The Membrane Paradigm.* New Haven: Yale University Press,
1986.

"Those Baffling Black Holes." *Time,* 4 September 1978, 50–59.

Tucker, Wallace, and Riccardo Giacconi. *The X-Ray Universe.* Cambridge,
MA: Harvard University Press, 1985.

Wade, C. M., and R. M. Hjellming. "Position and Identification of the Cygnus X-1 Radio Source." *Nature* 235 (1972): 271.

Wade, Richard A., et al. "A Sharpened Hα + [N II] Image of the Nebula
Surrounding Nova V1500 Cygni (1975)." *Astrophysical Journal* 102 (1991):
1738–41.

Wali, Kameshwar C. "Chandra: A Biographical Portrait." *Physics Today* 63
(December 2010): 38–43.

———. *Chandra: A Biography of S. Chandrasekhar.* Chicago: University of
Chicago Press, 1992.

———. "Placing Chandra's Work in Historical Context." *Physics Today* 64
(July 2011): 7, 9.

Welther, B. L. "The Discovery of Sirius B: A Case of Strategy or Serendipity?" *Journal of the American Association of Variable Star Observers* 16
(1987): 34.

Westfall, Richard S. *Never at Rest: A Biography of Isaac Newton.* Cambridge:
Cambridge University Press, 1980.

Wheeler, John Archibald. *A Journey into Gravity and Spacetime.* New York: Scientific American Library, 1990.

———. "The Lesson of the Black Hole." *Proceedings of the American Philosophical Society* 125 (February 1981): 25–37.

———. "Our Universe: The Known and the Unknown." *American Scientist* 56 (1968): 1–20.

———. "The Superdense Star and the Critical Nucleon Number." In *Gravitation and Relativity*, ed. Hong-Yee Chiu and William F. Hoffmann, 195–230. New York: W. A. Benjamin, 1964.

———. "The Universe in the Light of General Relativity." *Monist* 47 (1962): 40–76.

Wheeler, John Archibald, and Kenneth Ford. *Geons, Black Holes, and Quantum Foam.* New York: W. W. Norton, 1998.

Wheeler, J. Craig. *Cosmic Catastrophes: Exploding Stars, Black Holes, and Mapping the Universe.* Cambridge: Cambridge University Press, 2007.

Wolpert, Stanley. *A New History of India,* 8th ed. New York: Oxford University Press, 1997.

Zel'dovich, Ya. B. "The Fate of a Star and the Evolution of Gravitational Energy upon Accretion." *Soviet Physics Doklady* 9 (1964): 195.

Zel'dovich, Ya. B., and O. H. Guseynov. "Collapsed Stars in Binaries." *Astrophysical Journal* 144 (1966): 840–41.

Zel'dovich, Ya. B., and I. D. Novikov. "Gravitational Collapse ('Black Holes') and Searches for It." *Vestnik Akademii Nauk SSSR* 42 (February 1972): 16–20.

Zwicky, Fritz. *Morphological Astronomy.* Berlin: Springer, 1957.

———. "Die Rotverschiebung von extragalaktischen Nebeln [The redshift of extragalactic nebulae]." *Helvetica Physica Acta* 6 (1933): 110–27.

Acknowledgments

The idea for this book emerged as I was advising one of my students in the MIT Graduate Program in Science Writing on her master's thesis. Camille Carlisle, now an editor at *Sky and Telescope* magazine, was writing on the Event Horizon Telescope project, a worldwide effort to image the supermassive black hole that sits in the heart of the Milky Way galaxy. As we hovered together in my office over her manuscript, I began thinking: while many general-audience books have been written on both the latest theoretical models of black holes and the fascinating behaviors of black holes in this vast universe, few have solely focused on the tumultuous history of getting these bizarre celestial objects recognized in the first place. Moreover, here was a way to celebrate the upcoming hundredth anniversary of the general theory of relativity; black holes played a vital role in bringing Einstein's highest accomplishment back to the forefront of physics.

Many offered insight and advice as I carried out my research, including scientists, historians, journalists, and a number of people who played a part in the story. I must specifically thank David Cassidy, Frances Chambers, Hong-Yee Chiu, John Dicke, Robert Fuller, Karl Hufbauer, David Kaiser, Roy Kerr, Alan Lightman, Martin McHugh, Charles Misner, Roger Penrose, Joe Polchinski, Albert Rosenfeld, Virginia Trimble, and Barbara Welther. Werner Israel, in particular, provided invaluable guidance as I consulted him a number of times on various aspects of both the science and the history that he himself participated in. Also immensely helpful were the staff and archivists at both the American Philosophical Society in Philadelphia and the American Institute of Physics in College Park, Maryland.

As I carried out my explorations into the history of black holes, I included a few of the stories in my column, "Cosmic Background," in *Natural History* magazine, where editors Vittorio Maestro and Erin Espelie wielded their magic on my copy. In addition, summary descriptions in this book on Newton's accomplishments and the physics of relativity were partly drawn from my previous book *Einstein's Unfinished Symphony*, which focused on gravity-wave astronomy.

Deep appreciation goes to my literary agents Shannon O'Neill and Will Lippincott, at the Lippincott Massie McQuilkin agency, who found the perfect publishing home for my book at the Yale University Press. It was simply a delight to work with my editor, Joseph Calamia, who dove into my manuscript from the start with such joyous enthusiasm, keen insight, and journalistic wisdom. Wordsmith extraordinaire Laura Jones Dooley followed up with a copy edit that writers only dream of.

Heartfelt love and gratitude go to my husband, Steve Lowe, for his support, his patience, and his editorial advice over the course of this endeavor. Last, I can't forget to thank the one who was by my side (literally) throughout the writing process: my dog, Hubble, both a champ and a scamp.

Index

Note: Italic page numbers refer to illustrations.